The U.S. Steel Industry in Recurrent Crisis

ROBERT W. CRANDALL

The U.S. Steel Industry in Recurrent Crisis

Policy Options in a Competitive World

THE BROOKINGS INSTITUTION
Washington, D.C.

THE BROOKINGS INSTITUTION
1775 Massachusetts Avenue, N.W., Washington, D.C. 20036

Library of Congress Cataloging in Publication data:

Crandall, Robert W.
The U.S. steel industry in recurrent crisis.
Includes bibliographical references and index.
1. Steel industry and trade—United States.
I. Brookings Institution. II. Title.
HD9516.C7 338.4'7669142'0973 81-4642
ISBN 0-8157-1602-8 AACR2
ISBN 0-8157-1601-X (pbk.)

9 8 7 6 5 4 3 2 1

THE BROOKINGS INSTITUTION is an independent organization devoted to nonpartisan research, education, and publication in economics, government, foreign policy, and the social sciences generally. Its principal purposes are to aid in the development of sound public policies and to promote public understanding of issues of national importance.

The Institution was founded on December 8, 1927, to merge the activities of the Institute for Government Research, founded in 1916, the Institute of Economics, founded in 1922, and the Robert Brookings Graduate School of Economics and Government, founded in 1924.

The Board of Trustees is responsible for the general administration of the Institution, while the immediate direction of the policies, program, and staff is vested in the President, assisted by an advisory committee of the officers and staff. The by-laws of the Institution state: "It is the function of the Trustees to make possible the conduct of scientific research, and publication, under the most favorable conditions, and to safeguard the independence of the research staff in the pursuit of their studies and in the publication of the results of such studies. It is not a part of their function to determine, control, or influence the conduct of particular investigations or the conclusions reached."

The President bears final responsibility for the decision to publish a manuscript as a Brookings book. In reaching his judgment on the competence, accuracy, and objectivity of each study, the President is advised by the director of the appropriate research program and weighs the views of a panel of expert outside readers who report to him in confidence on the quality of the work. Publication of a work signifies that it is deemed a competent treatment worthy of public consideration but does not imply endorsement of conclusions or recommendations.

The Institution maintains its position of neutrality on issues of public policy in order to safeguard the intellectual freedom of the staff. Hence interpretations or conclusions in Brookings publications should be understood to be solely those of the authors and should not be attributed to the Institution, to its trustees, officers, or other staff members, or to the organizations that support its research.

Foreword

Twenty years ago the U.S. steel industry was the largest and most efficient in the world, but since then its fortunes have declined sharply. It not only has ceased to grow but has recently been retiring more capacity than it has been building. Throughout the 1960s and 1970s the industry was so unprofitable that it was unable to attract the capital to modernize its productive facilities. As a result, it enters the 1980s in a weakened state, subject to recurrent crises during cyclical downturns.

In this book Robert W. Crandall seeks to identify the root causes of the industry's decline, focusing on the diffusion of steelmaking technology throughout the world and on falling ore prices and shipping costs as well as on high U.S. wage rates. In analyzing various policies for stemming this decline, he questions the soundness of raising import barriers, concluding that trade protection for the steel industry is futile and costly to American consumers and steel-using industries. Moreover, he challenges the "national security" argument often used in defense of protectionism, as well as the idea that modernizing by building new plants will solve the steelmakers' problems. For these reasons his findings are likely to be sharply debated.

Although Crandall foresees a continued slow contraction, he believes that eventually a smaller, more efficient industry located around the Great Lakes will be better able to hold its own against the expansion of steel-producing capacity in the less developed countries, particularly in eastern Asia and Latin America. But he stresses that the U.S. industry cannot regain its position of the late 1950s unless American consumers are prepared to pay substantial premiums for products fabricated from steel.

Robert Crandall, a senior fellow in the Brookings Economic Studies program, came to the institution in 1978 after serving as deputy director of the Council on Wage and Price Stability, where he helped draft the Solomon Plan for the steel industry in 1977. His research for this study benefited from the assistance of Jeffrey Goldstein, Arthur Kupferman, Lewis Alexander, and Gregory Call. Useful comments and suggestions were received from Barry P. Bosworth, Donald Barnett, Hans Mueller, Kyoshi Kawahito, Jeff Vaughan, and Robert Z. Lawrence. Pamela Dobson and Lisa Saunders provided invaluable assistance in preparing the manuscript.

Tadd Fisher edited the manuscript, Penelope Harpold checked it for accuracy, and Patricia Foreman prepared the index.

The views expressed here are those of the author and should not be ascribed to the trustees, officers, or other staff members of the Brookings Institution.

BRUCE K. MACLAURY
President

February 1981
Washington, D.C.

Contents

Text Tables

Appendix Tables

Figures

Introduction

The year 1977 marked a major turning point for the U.S. steel industry. The industry's recovery from the deep 1974–75 recession was aborted by a sudden surge in imports and the price cutting associated with this surge. Two major producers announced plant closings and a third smaller company suspended production altogether. Net income for the industry fell to virtually zero. Rumors spread that the industry would close more plants unless the government quickly came to its assistance. Its cry was heeded.

The Trade Policy Crisis

With industry and the United Steelworkers of America applying the pressure, the Congressional Steel Caucus of more than 200 members was hastily formed. Protectionist legislation was readied. The White House was bombarded by complaints that the administration had failed to enforce the trade laws. The Carter administration responded by inviting the steel industry to bring formal trade cases to the Treasury Department so that "unfair" import competition could be dealt with according to the 1974 revision of the Antidumping Law. This led to a spate of dumping complaints against the European and Japanese exporters at a time when the United States and its trading partners were in the final stages of negotiating the Tokyo Round of tariff reductions under the General Agreement on Tariffs and Trade. Prosecution of the steel antidumping cases would surely have interfered with the successful consummation of the multilateral trade negotiations, and it would have added inflationary pressures at a time when inflation was accelerating dangerously.

Once the full implications of the new dumping suits were understood by the Carter administration, it became necessary to construct a set of policies to ease the pressure on the steel industry and its employees, induce the companies to withdraw or suspend their dumping complaints, quiet the congressional proponents of trade protection, and minimize the contribution to domestic inflation, which was rising to 7 percent and beyond. To accomplish these rather difficult tasks, the president's economic advisers sought out

Anthony Solomon, under secretary of the Treasury for monetary affairs. Solomon had been the architect of an earlier plan to provide unlegislated trade protection for the steel industry in the 1969–74 period. He was now asked to turn his attention from the falling dollar to the more precipitous apparent decline of the steel industry.

Solomon's response was swift. To ease the trade policy problem, he suggested establishing a set of reference, or "trigger," prices based on the cost of production in the world's most efficient steel industry—the Japanese industry. These prices would serve as prima facie benchmarks on which to base antidumping investigations under the new cost-of-production standard in the revised Antidumping Law. In addition smaller, struggling companies with mills in declining areas would be candidates for loan guarantees under Economic Development Administration programs at the Department of Commerce. The Treasury would reexamine its assessment of the useful economic life for depreciating steel assets. The Environmental Protection Agency would attempt to rationalize its policies toward the industry so as to reduce the burden of compliance costs without compromising environmental goals. Finally, a tripartite committee of labor, industry, and government officials would be established to formulate further policy options and to follow the progress of Solomon's plan.

The industry, the union, and Congress accepted the Solomon Plan with relatively little public complaint. The antidumping suits were suspended. Suggestions for further trade protection were abandoned. The industry and its employees settled in for a period of trial of the trigger price mechanism. Solomon had accepted his task in September. The first trigger prices were announced on January 3, 1978.

During this 1977–78 crisis period the European industry adopted the Davignon Plan (see chapter 2), which provided for domestic guidance prices, import reference prices, and the phased reduction of steelmaking capacity. The Japanese sharply reduced their exports to the European Economic Community and exercised restraint in selling to the American market. Thus the three major steel-producing regions had implicitly reached a set of arrangements to control the aggressive competition that had emerged in the world market. Export prices appeared to firm, and Japanese and U.S. steelmakers' profits rose in 1978–79, but the European industry continued to founder.

The crisis conditions of 1977–78 also led the developed steelmaking countries to form a Steel Committee within the Organisation for Economic Co-operation and Development. This committee has not produced any dramatic results in its first few years of existence, but it clearly offers the possibility for cooperation to substitute partially for competition as Europe, the United States, and Japan struggle to restore equilibrium in the world steel market.

A Chronic Problem

The events of 1977–78 were not unique. The U.S. steel industry has faced recurrent crises since the late 1950s. Imports began to surge in 1959 when steelmakers faced a long strike; throughout the 1960s they struggled with sluggish growth, rising imports, low profits, and repeated confrontations with government over pricing policies. Finally, in 1968 the import problem became so severe that the industry persuaded the government to negotiate import restrictions with Japanese and European exporters of steel.

Despite protectionism during the 1969–74 period, the industry did not rebound from the doldrums of the 1960s. Although in 1974 strong demand and the abandonment of price controls combined to produce a profitable year, the 1975 recession plunged the industry back into a depressed state—a condition from which it has not recovered. The slow growth of the 1960s has now been replaced by an actual reduction of total industry capacity. Environmental costs have risen sharply. Labor costs have been rising at a precipitous rate. After two major bouts of trade protection, the industry appears to be in a more precarious position than ever.

The Plan of the Book

What has caused the crisis in the U.S. steel industry? Why has the U.S. government suddenly begun to take an interest in "saving the steel industry"? In this book I attempt to place the U.S. industry's economic plight in perspective. The first two chapters trace the industry's decline from the preeminent role it enjoyed immediately after World War II and the rise of Japan to leading exporter in the world.

The third chapter provides a simple empirical model of the determinants of export prices, U.S. import prices, U.S. producers' prices, and U.S. market shares. Various theories of domestic and world market performance are tested, with particular emphasis on the determinants of relative price movements.

Chapter 4 is an analysis of current and future carbon steel production costs in various parts of the world. The relationship between labor rates, plant construction costs, raw material prices, and the location of future steel-producing assets is the central focus of the analysis.

Chapters 5, 6, and 7 constitute a detailed review of the rationale for trade protection and the actual outcome of the protection granted the U.S. industry in 1969–74 and 1978–79. The national security argument for maintaining a large domestic industry (satisfying 85 to 90 percent of U.S. requirements) is examined in depth. The social cost of the two recent bouts of protection—the Voluntary Restraint Agreements of 1969–74 and the trigger price mechanism of 1978–79—is measured in chapters 6 and 7.

Finally, the concluding chapter is an assessment of the future of the domestic steel industry in a competitive world market and is based on the assumption of relatively free trade and current technology. An attempt is made to predict the likely level of viable capacity, its location, and the labor force it will support.

I

An Overview of the U.S. Steel Industry

This monograph addresses the current problems of the U.S. iron and steel industry, problems that have been brewing for more than two decades. It is concerned mostly with issues of new-plant investment, import competition, and the appropriate government policy for preserving the necessary steel capacity for U.S. national security. It does not address the problems of the entire steel industry. Rather, it focuses only on what is known as basic carbon steel, which is produced by large integrated plants.

This focus on integrated carbon steel production was chosen because the basic carbon steel industry is the subject of most of the current policy debate concerning production efficiency, pricing, wage rates, and trade protection. Smaller carbon steel producers and "specialty" steel firms offer a different range of products, use a different technology, and are subject to different economic and competitive conditions. Conclusions based on the price behavior, cost conditions, and international competitiveness of the U.S. carbon steel industry cannot be extended to the specialty steel industry. This is not to say that there are no interesting policy problems involving firms in the latter category but only that these issues should be separated from those relevant to the integrated carbon steel industry.

Steel Production Technology

The steel industry may be defined as the set of establishments that produce finished steel mill products and semifinished slabs, blooms, and billets from iron ore, steel scrap, or both. This steel varies considerably in chemical composition, but the industry generally categorizes its output as carbon, alloy, or stainless steel. Carbon steels constitute the bulk of world (and U.S.) steel production. These steel products are used for construction, in automobiles, and even in containers. Alloy and stainless steels are specialty items serving a variety of requirements, such as corrosion resistance in different media or

Table 1-1. *U.S. Raw Steel Output by Grade, 1965, 1970, 1975, 1978*

	Output			
Year	*Carbon grade*	*Alloy grade*	*Stainless grade*	*Total*
		Thousands of net tons		
1965	116,651	13,318	1,493	131,462
1970	117,411	12,824	1,279	131,514
1975	100,360	15,171	1,111	116,642
1978	116,916	18,161	1,954	137,031
		Percent of total		
1965	88.8	10.1	1.1	100
1970	89.3	9.7	1.0	100
1975	86.0	13.0	1.0	100
1978	85.3	13.3	1.4	100

Source: *Annual Statistical Report: American Iron and Steel Institute, 1978* (Washington, D.C.: AISI, 1979), table 26, p. 55.

extra strength when lightweight, high-strength materials are required. In recent years alloy steels have begun to substitute for carbon grades in the U.S. output mix (see table 1-1), but carbon steel continues to dominate the specialty grades, accounting for 85 percent of total steel production.

The steel industry uses a number of different technologies. After World War I the open hearth process was used to produce 75 percent of the U.S. raw steel output, and the Bessemer process was used for most of the remainder.[1] By World War II the open hearth had assumed a dominant role, accounting for 88 to 90 percent of production, with electric arc furnaces and the Bessemer process dividing the remaining 10 to 12 percent. Following World War II, the basic oxygen (or Linz-Donawitz) process began to replace open hearths in the United States and Thomas converters in Europe. This basic oxygen furnace offered much shorter heat times and therefore required much less labor and capital per ton of output. As a result, it has gradually replaced the open hearth in the United States, as table 1-2 shows. At the same time, electric furnace production has grown substantially.

Part of the growth in electric furnaces may be attributed to the increasing availability of steel scrap in the United States due to the advent of the shredding of automobile hulks and the adoption of the basic oxygen furnace technology. Basic oxygen furnaces can accommodate less scrap than open hearths unless the scrap is preheated at a substantial cost penalty. Electric

1. Data on production by furnace types can be found in relevant editions of *Annual Statistical Report: American Iron and Steel Institute* (Washington, D.C.: AISI, 1947).

Table 1-2. *U.S. Raw Steel Output by Furnace Type, 1965, 1970, 1975, 1978*

	Output			
Year	Basic oxygen furnace	Open hearth furnace	Electric furnace	Total
		Thousands of net tons		
1965	22,879	94,193	13,804	130,876
1970	63,330	48,022	20,162	131,514
1975	71,801	22,161	22,680	116,642
1978	83,484	21,310	32,237	137,031
		Percent of total		
1965	17.5	72.0	10.5	100
1970	48.2	36.5	15.3	100
1975	61.6	19.0	19.4	100
1978	60.9	15.6	23.5	100

Source: *Annual Statistical Report: American Iron and Steel Institute, 1978* (Washington, D.C.: AISI, 1979), table 26, p. 55.

furnaces, on the other hand, use virtually 100 percent scrap (or directly reduced iron where natural gas is abundant) as a charge.

Another reason for increasing reliance on electric furnaces derives from enforcement of pollution laws against open hearths, blast furnaces, and coke ovens. Electric furnaces do not require hot metal; hence the entire complex of old coke ovens, blast furnaces, and open hearths can be replaced at much lower capital and environmental control costs with an electric furnace shop. Firms such as Bethlehem Steel Corporation (Johnstown, Pennsylvania) and Jones and Laughlin Steel Corporation (Pittsburgh, Pennsylvania) have availed themselves of this strategy in older plants faced with the twin pressures of obsolescence and environmental violations.[2]

The difference between electric furnace production and basic integrated steel production (using the basic oxygen furnace) can be seen in figure 1-1. The integrated facilities include iron ore yards, coal yards, coke ovens, and blast furnaces. Coke is produced from metallurgical coal to support the iron ore burden in the blast furnace. Limestone is added to the blast furnace, and the coke-lime-iron-ore combination is converted into impure hot metal (pig iron) and slag. The coke ovens, blast furnaces, and raw-material facilities constitute an important share of the total investment in an integrated steel plant.[3] In addition a sintering or pellet plant may be required to convert ores

2. *American Metal Market,* June 26, 1979, p. 4; and ibid., June 29, 1979, p. 1.
3. See chapter 4 for estimates of capital costs.

Figure 1-1. *Steel Production Processes*

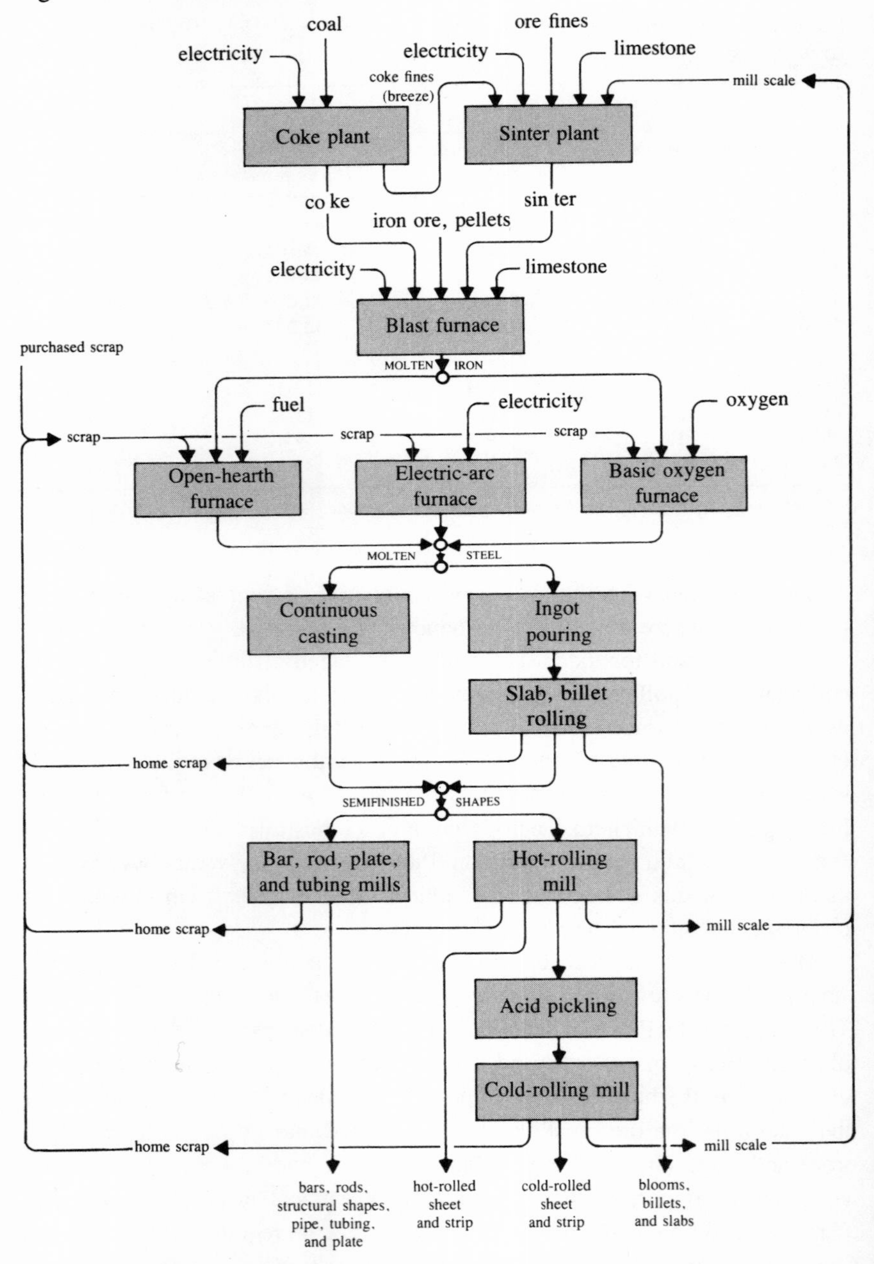

Source: U.S. Council on Wage and Price Stability, *Report to the President on Prices and Costs in the United States Steel Industry, 1977* (COWPS, October 1977), p. 14.

Note: A circle at a junction indicates alternatives.

into high-grade agglomerates suitable for blast furnace burdens. The pellet plants are typically near the mines producing lower-grade ores. The sintering plants (strands) are generally located at the steel plant site.

From the steelmaking furnace forward, there is very little difference between an integrated and nonintegrated mill. Steel may first be formed into ingots and then rolled into semifinished products or it may be cast into these semifinished forms directly through a fairly recent innovation, continuous casting. If the ingot-pouring route is chosen, a greater loss of product generally occurs since the semifinished forms are trimmed before they are fully rolled into final products.[4] The first solid form—either the ingot or the continuously cast slab or billet—is generally referred to as raw steel.

The semifinished shape is rolled or otherwise formed (forged, extruded) into final mill, or finished, steel products. These final products take a myriad of shapes and sizes, depending on their intended use. Sheet products (flat-rolled sheets of steel in various widths and thicknesses) are shipped to the automotive, container, and appliance industries. Plates (heavier flat-rolled sheets) are shipped to the construction, shipbuilding, industrial equipment, and railroad industries. Bar products (bars, reinforcing bars, and small structural shapes) are rolled from smaller cross sections, or billets, and are shipped to a large number of different users. Wire and tubular products are classified as mill products in the United States because major steel producers manufacture them, but elsewhere they are often defined as fabricated products. Wire is drawn from various grades of wire rod; pipes and tubes may be extruded from bars or rolled from sheets and plates and then welded. Finally, a number of miscellaneous products such as rails, track accessories, and even nails are called steel mill products in the United States.

The distribution of U.S. producers' shipments is displayed in table 1-3 along with major market classifications. Most attention in this study will be devoted to carbon sheet products, plates, structurals, and bars. Wire and tubular products are so heterogeneous that they cannot be easily characterized in an empirical model and indeed are often considered fabricated products in other countries. Moreover, the basic integrated carbon steel industry is dominated by flat-rolled and bar products. Wire rod and small bar products are produced by a vibrant, healthy mini-mill (electric-furnace) sector, while wire and tubular products are produced by a wide array of smaller firms that

4. This "yield loss" provides part of the reason why the yield from raw steel to finished steel is lower in the United States than in Japan, but its importance is often overstressed. Other technological differences, including greater computerization and larger ingot size, also contribute to the Japanese superiority in yield.

Table 1-3. *The Distribution of U.S. Domestic Steel Shipments by Product and Market Classification, 1978*
Millions of net tons

	Product					
Market classification	*Sheet and strip*[a]	*Bar products*	*Plate*	*Large structurals*	*All others*[b]	*All products*
Construction and contractors' materials	4.82	2.42	1.94	2.35	0.57	12.10
Automotive	17.12	2.98	0.24	0.07	0.49	20.90
Railroad transportation	0.54	0.26	0.83	0.34	1.57	3.54
Shipbuilding	0.07	0.03	0.64	0.09	0.01	0.84
Industrial machines	1.24	1.43	1.79	0.21	0.70	5.37
Electrical equipment	2.26	0.14	0.21	0.01	0.00	2.62
Appliances, utensils, cutlery	2.01	0.04	0.00	0.00	0.01	2.06
Containers and packaging	6.51	0.00	0.01	0.00	0.03	6.55
Steel service centers	9.04	1.94	1.79	1.22	0.38	14.37
All others	5.62	7.57	1.05	0.28	3.17	17.69
All market classifications	49.23	16.81	8.50	4.57	6.93	86.04

Source: *Annual Statistical Report: American Iron and Steel Institute, 1978* (Washington, D.C.: AISI, 1979), table 16, p. 35.
a. Includes tin plate and galvanized steel.
b. Excludes wire and tubular products.

compete successfully with the integrated companies. But the sheet and plate products are largely the domain of the major integrated firms.

The Size Distribution of Integrated Firms

The two facilities subject to the greatest economies of scale in the industry are the blast furnace and the hot strip mill. The Japanese have pioneered in the development of large blast furnaces, some of which are capable of producing 4 million tons or more of hot metal a year. Most of Japan's hot metal is produced by blast furnaces with an inner volume of 2,000 cubic meters or more (capable of producing 1.6 million tons or more of hot metal annually), while only five U.S. blast furnaces are this large. The minimum efficient scale is probably between 2,500 and 3,500 cubic meters, producing between 2 and 3 million tons a year.[5] Unfortunately blast furnaces and steel furnaces must be shut down periodically for relining. This creates a major logistic problem during the months in which the furnace is not available. The usual solution to this problem is to have two or three blast furnaces of less than full minimum efficient scale serving three or more basic oxygen furnaces so that at least one or two of each can be operating at a time.

The second source of large scale economies is the hot strip mill used to roll all slabs into the first sheet form in producing hot- or cold-rolled sheet, galvanized sheet, and tin plate. This mill has a minimum efficient scale (when fully continuous) of between 3 and 4 million tons.[6] It is not subject to lengthy periods of idleness for maintenance; therefore a plant capable of producing 4 million tons of hot bands is likely to be sufficiently large to exhaust rolling economies.

Given the minimum efficient scale in blast furnaces and the indivisibility constraints in planning a combination of blast furnaces, basic oxygen furnaces, breaking mills, and finishing facilities, the minimum efficient scale of a new integrated "greenfield" plant is somewhere between 6 and 7 million raw tons.[7] This can be achieved with two or three blast furnaces, three or four basic oxygen furnaces, a combination of continuous casting and ingot pouring, one large hot strip mill, and a combination of other finishing facilities such as a plate mill, a cold-rolling mill, a tinning line (for tin plate), and a galvanizing line. In addition a combination of billet and bar mills can be included either as a substitute for the plate mill or as a complement to the other facilities in a larger plant, extending to 9 million or 10 million tons.

Since the U.S. industry has approximately 153 million tons of raw steel capacity (or "capability"),[8] the discussion above would seem to suggest that

5. For a discussion of economies of scale in blast furnaces see Myles G. Boylan, Jr., *Economic Effects of Scale Increases in the Steel Industry: The Case of U.S. Blast Furnaces* (Praeger, 1975); U.S. Federal Trade Commission, *The United States Steel Industry and Its International Rivals: Trends and Factors Determining International Competitiveness* (Government Printing Office, 1978), chap. 7; and Julian Szekely, "Blast Furnace Technology," in Julian Szekely, ed., *The Steel Industry and the Energy Crisis,* Proceedings of the Fourth C. C. Twinas Memorial Conference (New York: Marcel Dekker, 1973). Minimum efficient scale is the annual rate of output at which unit costs achieve their minimum.

6. T. J. Ess, *The Hot Strip Mill: Generation II* (Pittsburgh, Pa.: Association of Iron and Steel Engineers, 1970).

7. "Greenfield plant" refers to a new plant on a new site as opposed to a refurbished plant or new facilities on an old "brownfield" site.

8. *Annual Statistical Report: American Iron and Steel Institute, 1979* (Washington, D.C.: AISI, 1979), table 27, p. 55. Since 1979 a number of mills have been closed; therefore total industry capacity is somewhat lower than the 155.3 million tons reported for 1979.

Table 1-4. *Raw Steel Capacity of Major U.S. Integrated Steel Producers, 1978*
Millions of net tons per year

Company		*Raw steel capacity*
United States Steel Corp.		38.0
Bethlehem Steel Corp.		21.5
National Steel Corp.		13.2
LTV (includes Jones and Laughlin Steel Corp. and Youngstown Sheet and Tube Co.)		13.4
Republic Steel Corp.		11.2
Armco, Inc.		9.9
Inland Steel Co.		8.6
Wheeling-Pittsburgh Steel Corp.		4.4
Total, eight largest producers		120.2
Kaiser Steel Corp.		3.2
Ford Motor Co., Steel Division		3.8
McLouth Steel Corp.		2.0
CF&I Steel Corp.		1.9
Wisconsin Steel (suspended production in 1980)		1.4
Interlake, Inc.		1.4
Sharon Steel Corp.		1.4
Passco (Pacific States Steel Corp.)		0.4
Total (sixteen largest integrated producers)		135.7
Industry capability		157.9
Addendum		
Percent of industry capacity produced by:		
Four largest integrated producers	54.6	
Eight largest integrated producers	76.2	
Sixteen largest integrated producers	86.0	

Sources: Annual reports of individual companies to the U.S. Securities and Exchange Commission; company listings in *Iron and Steel Works of the World,* 7th ed. (New York: Metal Bulletin Books, 1978). Industry capacity is from *Annual Statistical Report: American Iron and Steel Institute, 1978* (Washington, D.C.: AISI, 1979), table 26, p. 55.

there is room for no more than twenty-five integrated plants or even fewer, given the share of output captured by electric-furnace producers. Yet the largest five firms in the industry alone have twenty-seven integrated plants and a number of smaller electric-furnace and finishing facilities.

There are only seven firms in the United States with a total capacity equal to one new plant of minimum efficient scale. The top eight firms account for 76 percent of the industry's capacity (table 1-4), but their share of output has actually been falling for twenty-five years. The merger of Jones and Laughlin and Youngstown Sheet and Tube in 1978 has probably altered this trend at least temporarily.

Table 1-5. *Capacity of Japanese Integrated Steel Works, 1978*
Millions of net tons per year

Company	*Plant*	*Capacity*
Nippon Steel Corp.	Yawata	9.73
	Muroran	4.82
	Kamaishi	1.53
	Hirohata	4.44
	Nagoya	8.27
	Sakai	5.20
	Kimitsu	10.45
	Oita	9.26
KKK-Nippon Kokan KK	Keihin (Ohgishima)	6.06
	Fukuyama	17.64
Sumitomo Metal Industries, Ltd.	Wakayama	10.19
	Kashima	9.92
	Kokura	3.44
Kobe Steel, Ltd.	Kakogawa	7.07
	Kobe	2.83
	Amagasaki	0.77
Kawasaki Steel Corp.	Chiba	10.47
	Mizushima	13.31
Nisshin Steel Co., Ltd.	Kure	5.93
Total integrated works		141.35
Average capacity		7.44

Source: Based on information in *Iron and Steel Works of the World*, 7th ed. (New York: Metal Bulletin Books, 1978), individual company listings.

It is instructive to compare the American and Japanese industries. The Japanese industry has been built almost entirely since World War II, but the U.S. firms built most of their facilities before 1945. Only two new integrated major plants have been built in the United States since 1950: United States Steel's Fairless Works in eastern Pennsylvania and Bethlehem's Burns Harbor plant on Lake Michigan.[9] The average size of an integrated plant in the United States is substantially less than 3 million tons. The five largest Japanese producers have more than 80 percent of the Japanese capacity, most of which is in their seventeen integrated plants, and the average size of these plants is more than 7 million net tons (see table 1-5). Only three integrated plants among these seventeen are as small as the average U.S. integrated works.

There are a number of good reasons for this disparity in size distribution

9. Other greenfield sites have included Portage, Indiana, and Conneaut, Ohio (and Pennsylvania). In Portage, National has built rolling mills, but it has not followed through with earlier plans to build steelmaking and ironmaking facilities there. Conneaut remains a potential site for an integrated U.S. Steel plant (see chapter 4).

between the two countries. First, the minimum efficient scale in production has increased since World War II. Second, retrofitting old plants to new technology is difficult when larger scale and continuity of operations are required. Space constraints preclude expansion of many plants and simple economics argue against complete suspension of production while new, larger facilities replace the older generation of facilities at the same plant sites. Third, the geographic market served by many U.S. plants is often a constraint on expansion. Finally, the incremental cost of producing a ton of steel in old, smaller facilities is generally less than the average cost of production in a new, larger, "efficient" works.[10] Therefore, the U.S. industry will not be moving toward the average size of Japanese plants any time soon.

The Smaller Nonintegrated U.S. Companies

The electric furnace has grown in importance almost continuously since World War II. In the early 1940s electric furnaces produced only 5 percent of the country's steel.[11] At present they account for nearly 25 percent. A major contributor to this rise has been the mini-mill, a small producer of wire rods and bar products, generally of carbon steel. These mills use electric furnaces and virtually 100 percent scrap in their metallics charge. Few mini-mills are as large as 1 million tons in annual capacity; the largest are generally in the range of 500,000 to 700,000 tons. A number of other specialized electric-furnace companies produce steel plates and some sheet products, but most electric-furnace mills compete with the integrated firms in only a narrow range of product lines: wire rods, bars, bar shapes, and (to a small extent) plates. They are not a major force in sheet or large structural markets.[12]

In 1978 the electric-furnace companies (including mini-mills) had approximately 17.8 million tons of capacity, over 10 percent of the industry total (see table 1-6). Their plants are spread across the country to take advantage of locally produced scrap and proximity to various geographic markets. In general their fortunes are much less susceptible to cyclical variations in demand than are those of their larger brethren. When demand falls, scrap prices decline markedly and the electric-furnace companies' product prices

10. See chapter 4 for a fuller discussion of this point.

11. *Annual Statistical Report, 1946,* p. 32.

12. For a description of the growth of mini-mills, see "Mini-Mill Roundup," *33: McGraw-Hill's Magazine of Metals Producing,* vol. 12 (July 1974), pp. 36–43; and Darwin I. Brown, "Mini and Medium Steel Plants of North America," *Iron and Steel Engineer,* vol. 52 (November 1975), pp. 1–29.

Table 1-6. *Capacity of Major U.S. Electric-Furnace Steel Producers, 1978*
Millions of net tons per year

Company	*Raw steel capacity*
Northwestern Steel and Wire Co.	2.70
Georgetown Steel Corp. (2 plants)	1.20
Nucor Corp. (3 plants)	0.95
Laclede Steel Co.	0.84
Lukens Steel Co.	0.79
Atlantic Steel Co. (2 plants)	0.70
Penn-Dixie Steel Corp.	0.65[a]
Keystone Steel and Wire (a division of Keystone Consolidated Industries, Inc.)	0.65
North Star Steel Co. (2 plants)	0.60
Connors Steel Co. (2 plants)	0.58
Ceco Corp.	0.57
Phoenix Steel Corp.	0.50[a]
Raritan River Steel Co.	0.50
Oregon Steel Mills (division of Gilmore Steel Corp.)	0.50
33 others	6.07
Total	17.80[b]

Sources: *Iron and Steel Works of the World,* 7th ed. (New York: Metal Bulletin Books, 1978), individual company listings; and "Steel, USA: Into the 80's," *IISS Commentary,* vol. 9 (January 1980).
a. Author's estimate.
b. Estimated by extrapolating between IISS estimates for 1977 and 1980 and adding the capacities of Lukens, Phoenix, and Oregon.

follow. As a result, these companies are better able to compete with large firms and importers for market share, and they are not as active in seeking trade protection as the integrated firms. During booms, however, scrap prices soar, and the companies are frequently heard to complain about excessive U.S. scrap exports.

The Growth in the U.S. Steel Market

In an advanced economy such as that of the United States steel demand does not grow at a very rapid rate. In the three years preceding the 1958 recession U.S. apparent supply of steel averaged 79.2 million tons.[13] In 1978–79 it averaged 112.5 million tons. This represents an average annual

13. Apparent supply is equal to production plus imports less exports. It excludes inventory adjustments.

Figure 1-2. *U.S. Raw Steel Production and Industrial Production, 1956–78*

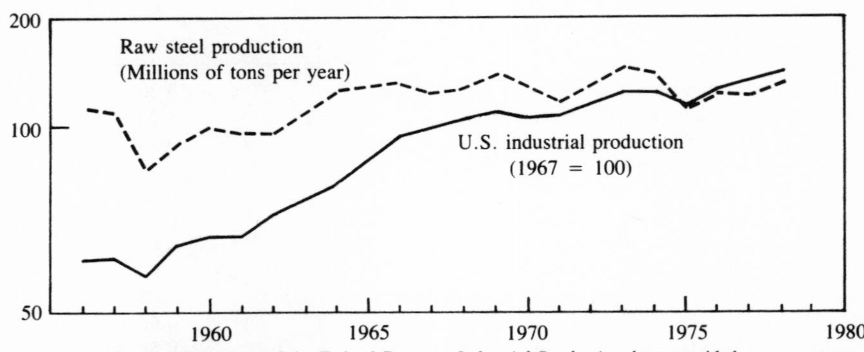

Sources: Board of Governors of the Federal Reserve, *Industrial Production* data, provided on computer tape by Data Resources, Inc.; see also raw steel data in annual statistical reports of the American Iron and Steel Institute.

rate of growth of approximately 1.6 percent. With rising imports, U.S. production has grown even more slowly. Since 1955–57 U.S. producers' shipments have grown by only 0.7 percent a year. Even if the 1956–74 period is used as the basis for estimating the long-term growth trend, U.S. shipments have grown by no more than 1.5 percent a year. By contrast, total industrial production grew by 4.1 percent a year in the same period. (See figure 1-2.)

Much of the growth in steel demand in past years has originated in the construction and automotive industries. With the interstate highway system largely built, new materials substituting for steel in construction, and a decline in the average size of automobiles, there has been a decided slowing of demand. Automotive industries accounted for only 16 percent of the industry's shipments in the 1950s. This share rose to 19 percent in the first half of the 1970s and to as high as 24 percent in 1976–77. With the reduction in the size of cars and the substitution of lightweight materials for steel, this share may be expected to fall substantially. Shipments to construction industries began declining in relative importance in the 1960s, and this decline persisted in the 1970s.

Because of relatively sluggish growth and strong import pressures, U.S. producers have operated with considerable average excess capacity over the past twenty years (table 1-7). This excess capacity has obviously restrained expansion plans and retarded productivity growth in the industry. As a result, U.S. producers are perceived to be sluggish and inefficient, but as chapters 4 and 6 will demonstrate, they have had precious little opportunity to build

Table 1-7. *Average Capacity, or "Capability," Utilization in the Steel Industry, 1956–78*

Period	Average utilization rate (percent)
1956–60	74.5
1961–65	77.2
1966–70	87.7
1971–75	85.2
1976–78	82.0

Source: U.S. Council on Wage and Price Stability, *Report to the President on Prices and Costs in the United States Steel Industry, 1977* (COWPS, October 1977), app. table 27, p. 145; and various issues of *Annual Statistical Report: American Iron and Steel Institute*.

new facilities. Rather, the integrated carbon steel producers have faced the unenviable task of keeping older facilities operating as well as possible and swallowing the inevitable capital losses.

II

The Origins of the Steel Problem

The steel industry has provided a "problem" for government policymakers since the Truman administration. During the late 1940s President Truman, in what turned out to be an erroneous forecast, feared that the industry would not increase capacity sufficiently rapidly to accommodate the increasing demand for steel. In his 1949 State of the Union Address he criticized the industry for this failure to expand.[1] Later, in 1951–52, he found himself trapped in a much more ticklish situation, one that exploded into a major government-industry confrontation. Upset at the industry's failure to settle with the United Steelworkers of America without governmental assurances of a large price increase, Truman attempted to seize the industry and to operate it temporarily. He was thwarted by a Supreme Court decision, but the industry-government confrontation had begun. It was given a major new impetus in 1962 by President Kennedy, who assailed the steel industry for raising prices after what he perceived to be a modest wage settlement, and has continued sporadically throughout the 1970s.

The Post–World War II Adjustment

Immediately after World War II the steel industry began a process of wage negotiations and price increases, dubbed the annual "rites of spring" by Adelman.[2] Each year, after settling for large wage increases, steel producers would announce major price increases that exceeded the increase in unit labor costs. According to the Adelman hypothesis, this annual ritual was required to coordinate price increases in an oligopolistic industry without running afoul of the antitrust laws. Between 1947 and 1957 the industry settled for increases

1. "Annual Message to the Congress on the State of the Union, January 5, 1949," *Public Papers of the Presidents of the United States: Harry S. Truman, 1949* (Government Printing Office, 1964), pp. 3–4. See Henry W. Broude, *Steel Decisions and the National Economy* (Yale University Press, 1963), for a discussion of the history of governmental intervention in steel pricing since World War II.

2. M. A. Adelman, "Steel, Administered Prices, and Inflation," *Quarterly Journal of Economics*, vol. 75 (February 1961), pp. 16–40.

Table 2-1. *Price of Steel, Iron Ore, and Steel Labor, 1947–57*

Year	*BLS price index for steel mill products, including alloy and stainless (1957–59 = 100)*	*Price of iron ore (dollars per net ton)*	*Total compensation per hour for hourly workers in the steel industry (dollars)*	*Unit labor costs (dollars per raw ton)*	*BLS wholesale price index, all commodities (1957–59 = 100)*
1947	45.5	5.55	1.95	29.02	81.2
1948	52.0	6.20	2.14	31.96	87.9
1949	56.4	7.20	2.19	33.35	83.5
1950	59.4	7.70	2.21	32.55	86.8
1951	64.0	8.30	2.47	36.40	96.7
1952	65.4	9.05	2.79	40.65	94.0
1953	70.5	9.90	2.89	40.12	92.7
1954	73.8	9.90	2.95	44.03	92.9
1955	77.2	10.10	3.08	40.25	93.2
1956	83.8	10.85	3.37	44.11	96.2
1957	91.8	11.45	3.76	48.90	99.0
Addendum					
Annual rate of change (percent)	7.0	7.2	6.6	5.2	2.0

Source: Richard Mancke, "The American Iron Ore and Steel Industries: Two Essays" (Ph.D. dissertation, Massachusetts Institute of Technology, 1968), tables 1-3 and 2-1, pp. 101 and 153.
BLS = Bureau of Labor Statistics.

in hourly compensation that averaged 6.6 percent a year while raising prices by 7 percent annually. (See table 2-1.) In the same period the average hourly wage in manufacturing was rising by only 5.2 percent. But the recession of 1958 and ominous developments in world steel markets presaging an increase in import competition led the industry to suspend the spring rites in 1959. Instead, the firms took a very long strike—that began in July 1959 and was not finally settled until January 1960—to attempt to reduce the rate of wage increase in the industry.

Whether the Adelman hypothesis is correct or not, the industry did increase its prices much more rapidly than other producer prices in the decade following World War II, as table 2-1 demonstrates. This might have been an adjustment to a competitive price equilibrium that was much higher than the prices World War II price authorities would allow, or it might have been a march toward joint-profit maximization. Unfortunately it is very difficult to determine which hypothesis fits the facts because, immediately after the plateau of 1958–59 was reached, the industry began to be buffeted by the winds of technological and market change. Foreign producers had begun to install modern equipment in the 1950s and were poised to compete in the U.S. market as the industry encountered its 1959 strike. The ensuing twenty years have markedly altered the 1957 "equilibrium."

An important part of the postwar adjustment involved the market for iron ore. Mancke explains that the iron ore industry emerged from World War II with its prices at their 1940 levels, despite widespread fears the U.S. deposits were in danger of being exhausted.[3] The industry started to react to these fears. In 1946 it began to raise its prices sharply to retard the exhaustion of the Mesabi range deposits. Non-Bessemer Mesabi ore rose from $4.55 in mid-1946 to $5.55 in 1947 and to $11.45 by 1957 after having remained virtually constant for twenty years (table 2-1).[4] This 7.2 percent annual rate of increase exceeded even the increase in the cost of labor, but much of it redounded to the owners of the deposits, notably, the steel companies themselves.

The fear of impending shortages led the U.S. industry and others to embark on a set of policies including some that would later have disastrous consequences for the steel industry. Processes were sought for using the Mesabi taconite ores. Major new processes for beneficiating low-grade ores were developed. New sources of iron ore were sought in Venezuela, Canada, and other countries. Larger, lower-cost bulk ocean liners were developed for shipping iron ore.

The result of these developments was a sharp increase in the available supply of U.S. and foreign ores in the 1960s. The Lake Erie price of U.S. iron ore would rise no longer. Imports of ore would begin flowing into the United States and into other countries from new deposits discovered in Venezuela, Australia, Canada, and elsewhere. The U.S. steel industry's advantage of owning its own, proximate, low-cost supply of iron ore would simply vanish.

The Painful 1960s

The U.S. steel industry never recovered from the recession of 1958 and the strike that followed in 1959. The opening of the world's iron ore market, reflected by the increase in U.S. imports of iron ore from less than 10 million tons a year in the Korean War period to 35 million tons in 1959–60 despite flat or slightly declining steel production between the two periods, provides

3. Richard Mancke, "The American Iron Ore and Steel Industries: Two Essays" (Ph.D. dissertation, Massachusetts Institute of Technology, 1968), table 1-2, pp. 37–38.

4. For ore prices before 1947, see ibid., table 1-3, p. 101.

Table 2-2. *Iron Ore and Metallurgical Coal Prices in the United States and Japan, 1956–76*
Dollars per net ton, including insurance and freight

	Coking coal		*Iron ore*	
Year	*United States*	*Japan*	*United States*	*Japan*
1956	9.85	22.14	9.63	16.69
1957	10.77	26.22	10.42	19.69
1958	10.48	19.31	10.61	14.70
1959	10.50	16.33	10.80	12.70
1960	10.56	15.63	11.15	12.88
1961	9.83	15.50	11.78	12.88
1962	9.70	15.35	11.60	12.97
1963	9.35	14.74	11.67	12.32
1964	9.85	14.43	11.88	12.21
1965	9.65	14.27	11.80	12.17
1966	9.82	14.41	11.74	11.91
1967	10.33	14.22	11.91	11.49
1968	10.59	14.40	12.31	11.09
1969	10.76	14.82	12.42	10.56
1970	12.27	18.29	13.05	10.74
1971	15.27	19.41	14.12	10.51
1972	17.68	19.87	15.04	10.37
1973	19.79	21.61	15.48	11.12
1974	34.22	40.71	19.62	13.26
1975	52.66	50.82	23.99	15.15
1976	56.04	53.60	27.62	15.81

Source: Based on data in U.S. Federal Trade Commission, *The United States Steel Industry and Its International Rivals: Trends and Factors Determining International Competitiveness* (Government Printing Office, 1978), table 3.3, p. 117.

the most dramatic index of change. More important, the price of iron ore began to fall in world markets. The Japanese, who were just beginning a major expansion of their steel industry in the late 1950s, were rewarded by declining iron ore and metallurgical coal prices. This decline began in 1958 and continued until 1970 (table 2-2). Not surprisingly, the U.S. price of iron ore flattened out completely in 1958.

The concentration of iron ore shipments in the world has declined markedly since the 1950s (table 2-3), and the concentration of metallurgical coal output has declined marginally. These developments, combined with lower shipping costs, have allowed virtually any country with good port facilities to obtain its basic raw materials as cheaply as the U.S. steel industry. Some countries, such as Japan, have been able to buy these basic materials at lower costs than their U.S. rivals can.

Table 2-3. *Worldwide Distribution of Iron Ore and Agglomerates Production, Selected Years*

Producer	*Percent of world production*				
	1954–58 average	*1962*	*1966*	*1970*	*1976*
United States	24.2	14.4	14.4	11.9	9.1
Canada	4.0	4.9	5.8	6.2	6.4
European Economic Community[a]	24.7	19.8	13.6	10.5	6.1
USSR	20.4	25.2	25.2	25.4	26.7
Brazil	1.1	2.1	3.7	5.2	10.3
Chile	0.7	1.6	1.9	1.5	1.2
Venezuela	2.9	2.6	2.8	2.9	2.0
Liberia	0.5	0.7	2.7	3.1	2.1
India	1.3	3.7	4.1	4.1	4.8
South Africa	0.5	0.9	1.2	1.0	1.7
Mauritania	0.0	0.2	1.1	1.2	1.1
People's Republic of China	3.4	5.9	6.3	5.9	7.3
Australia	1.0	1.0	1.7	6.7	10.4
Others	15.3	17.0	15.5	14.4	10.8
Total	100.0	100.0	100.0	100.0	100.0
United States, USSR, and European Economic Community	69.3	59.4	53.2	47.8	41.9

Source: Based on data in U.S. Department of the Interior, Bureau of Mines, *Minerals Yearbook, 1961*, vol. 1: *Metals and Minerals* (Government Printing Office, 1962); and ibid., 1963–77.

a. Includes Belgium, Denmark, France, West Germany, Ireland, Italy, Luxembourg, Netherlands, and United Kingdom.

The steel strike of 1959 invited the first rush of imports into the United States. Imports of steel mill products had averaged less than 1.5 million net tons from 1950 through 1958. Suddenly they rose to 4.4 million tons in 1959 (table 2-4). They were never again to recede to their levels of the early 1950s. Instead, they would rise in discrete jumps until 1971, usually increasing sharply in a year of steel labor negotiations and sometimes falling slightly in the subsequent year. By 1971 imports had reached 18.3 million tons, or 18 percent of U.S. apparent supply.

World exports increased dramatically during the 1960s, spurred by the decline in raw material and shipping costs and aided by the rapid improvements in productivity the Japanese and other exporters were obtaining from investment in modern facilities. For the three years preceding the 1958 recession, world exports averaged 13 percent of world production. One decade later, they were 17 percent of production and in 1974 were 24 percent of production. U.S. exports fell gradually during the period as Japanese and European exports soared. In 1957 the Japanese exported about 1 million tons of steel products; by 1976 they were exporting nearly 40 million tons (table 2-5). The rest of the world—other than the European Economic Community (EEC), the USSR, and the United States—also increased its steel exports markedly. By 1978 exports from these nations were greater than total world exports on the eve of the 1958 recession (table 2-6). The world steel industry was becoming much more competitive.

Although there are many reasons for the decline in the U.S. position in the world steel market, three are predominant: raw material prices, shipping costs, and new technology. The decline in raw material prices to an emerging steel exporter such as Japan are documented in table 2-2. The effects of these declines on unit production costs were dramatic: they lowered the costs of producing steel by approximately $25 a net finished ton from 1957 to 1967. Given that the average price of steel (excluding wire and tubular products) was $140 a net ton in the United States in 1967, this is a dramatic decline in the face of relatively constant U.S. coal and iron ore costs per ton of output.

Part of this decline in material costs can be traced to lower shipping costs. For instance, the cost of shipping iron ore from Brazil to Japan fell by 60 percent from 1957 to 1968.[5] The lower costs of shipping obviously facilitated exports to distant markets. Ironically, the United States and Canada aided the exporters substantially by building the St. Lawrence Seaway, linking the bulk of the U.S. market to Japanese and European trade.

Finally, the diffusion of new technology aided the new steel exporters in competing with the older, established firms in the United States. With a very low wage rate, Japan was able to compete in world markets even though its labor productivity was less than half of that achieved by the U.S. firms in the early 1960s. Subsequently Japanese wage rates increased as the Japanese industry's rate of productivity growth accelerated rapidly.

Much of this rapid rise in productivity can be traced to the program of building large new integrated mills at coastal locations. It is instructive to look at the rate of increase in Japanese capacity in the decade of the 1960s. The entire 1958 output of the Japanese industry was equal to the 1978 capacity of its second biggest mill, Mitsushima. Raw steel output in Japan increased sevenfold from 1958 through 1970, from 13.3 million tons to 102.9 million net tons of raw steel (table 2-5). As the Japanese adopted the newest steelmaking technology (the basic oxygen furnace), pioneered in large blast furnaces, and forged ahead rapidly with continuous casting, their labor pro-

5. Data on charter rates are available in *Chartering Annual* (New York: Maritime Research, 1957) and subsequent issues through 1969.

Table 2-4. *U.S. Capacity, Production, and Imports and World Production and Exports, 1956–78*
Millions of net tons

Year	United States				World		
	Raw steel capacity[a]	Raw steel production	Steel mill products: Imports	Steel mill products: Exports	Raw steel production	Exports (raw steel equivalent)	Percent exported
1956	129.9	115.2	1.3	4.2	313.4	39.5	12.6
1957	132.9	112.7	1.2	5.2	323.5	44.1	13.6
1958	136.3	85.3	1.7	2.7	302.6	41.9	13.8
1959	139.8	93.4	4.4[b]	1.7	337.6	46.4	13.7
1960	142.8	99.3	3.4	3.0	380.8	58.1	15.3
1961	143.5	98.0	3.2	2.0	390.5	57.7	14.8
1962	144.7	98.3	4.1[b]	2.0	395.4	61.7	15.6
1963	145.9	109.3	5.5	2.2	424.1	66.1	15.6
1964	147.5	127.1	6.4	3.4	478.6	76.4	16.0
1965	148.2	131.5	10.4[b]	2.5	503.8	86.5	17.2
1966	149.4	134.1	10.8	1.7	521.1	86.4	16.6
1967	150.6	127.2	11.5	1.7	547.8	94.4	17.2
1968	152.2	131.5	18.0[b]	2.2	583.9	109.1	18.7
1969	152.8	141.3	14.0	5.2	633.4	120.4	19.0

1970	153.8	131.5	13.4	7.1	656.2	129.4	19.7
1971	154.8	120.4	18.3[b]	2.8	642.0	138.2	21.5
1972	156.2	133.2	17.7	2.9	694.6	146.6	21.1
1973	156.7	150.8	15.1	4.1	769.5	161.8	21.0
1974	157.0	145.7	16.0[b]	5.8	781.2	187.2	24.0
1975	157.4	116.6	12.0	3.0	711.9	163.1	22.9
1976	157.7	128.0	14.3	2.7	746.0	179.7	24.1
1977	158.1	125.3	19.3[b]	2.0	744.5	179.7	24.1
1978	156.0	137.0	21.1	2.4	790.6	196.3	24.8
Addendum							
Annual percent change							
1959–69	0.9	4.1	11.6	11.2	6.3	9.5	...
1969–78	0.3	−0.3	4.6	−8.6	2.5	5.4	...
1959–78	0.6	2.0	8.3	1.8	4.5	7.6	...

Sources: *Annual Statistical Report: American Iron and Steel Institute*, selected years; and International Iron and Steel Institute, *World Steel in Figures, 1978* (Brussels: IISI, 1978); and ibid., *1980*, p. 14.

a. Capacity estimates are based on peak production months according to D. F. Barnett, "The Canadian Steel Industry in a Competitive World Environment" (Ottawa: Resource Industries and Construction Branch, Industry, Trade and Commerce, 1977), vol. 2, paper 14. The most recent peak month is May 1979: 12,789 net tons. The estimated 1979 capacity is assumed to be 12 times 12,789, or 153.5 million tons. Since the industry closed 3 percent of its capacity between 1977 and 1979, 1977 capacity is estimated to be 158.1 million tons. The annual estimates from 1974 through 1976 are interpolations and constant annual growth rates between the 1973 and 1977 estimates.

b. Year of expiration of labor contract.

Table 2-5. *Japanese Production, Capacity, Apparent Consumption, and Exports of Steel, 1956–78*
Steel amounts in millions of net tons

	Crude steel			*Finished steel*	
Year	*Production*	*Capacity*	*Apparent consumption*	*Exports*	*Man-hours per net ton*
1956	12.2	13.7	10.8	1.3	55.0
1957	13.9	16.3	14.2	1.0	51.6
1958	13.3	19.4	11.1	1.7	56.0
1959	18.3	23.3	16.3	1.7	44.8
1960	24.4	27.7	21.3	2.5	39.1
1961	31.2	33.1	28.1	2.5	34.2
1962	30.4	37.8	24.8	4.2	34.8
1963	34.7	42.0	26.9	5.8	32.2
1964	43.9	47.6	34.1	7.2	25.4
1965	45.4	54.3	31.4	10.5	24.2
1966	52.7	62.4	39.3[a]	10.4	20.4
1967	68.5	74.1	56.5	9.6	17.4
1968	73.7	85.4	55.9	14.1	16.1
1969	90.5	98.8	68.8	17.2	13.7
1970	102.9	114.3	78.4	19.4	12.4
1971	97.6	121.4	66.5	25.6	12.8
1972	106.8	131.0	78.3	23.0	11.7
1973	131.5	142.3	98.5	27.3	9.3
1974	129.1	154.5	87.1	35.5	9.1
1975	112.8	165.3	75.0	31.9	9.2
1976	118.4	166.4	71.9	39.7	9.1
1977	112.9	167.3	69.7	37.8	9.3[b]
1978	112.6	166.8	73.5	34.5	8.9[b]
Addendum					
Average annual growth rate (percent)	10.1	11.4	8.7	14.9	−8.3

Sources: Kimiro Suzuki and Tudor Mills, "Growth of Steel-Making Capacity in the 1980s," in Organisation for Economic Co-operation and Development, *Steel in the 80s: Paris Symposium, February 1980* (Paris: OECD, 1980), pp. 97–112; *Monthly Report of the Iron and Steel Statistics* (Tokyo: Japan Iron and Steel Federation), various issues; U.S. Council on Wage and Price Stability, *Report to the President on Prices and Costs in the United States Steel Industry* (COWPS, October 1977), pp. 107, 111; and U.S. Federal Trade Commission, *The United States Steel Industry and Its International Rivals: Trends and Factors Determining International Competitiveness* (Government Printing Office, 1978), pp. 238, 476.

a. For 1966 and following years, apparent consumption was calculated using conversion factors adopted by the Steel Committee of the UN Economic Commission for Europe.

b. Estimated.

Table 2-6. *Leading Exporters of Steel, 1978*

Country	*Exports (millions of finished net tons)*	*Share of world exports (percent)*
Japan	34.1	22.6
West Germany	20.5	13.6
Belgium/Luxembourg	14.7	9.7
France	12.0	7.9
USSR	7.2	4.8
Italy	9.1	6.0
Netherlands	5.1	3.4
Eastern European countries	13.0	8.6
South Korea	1.8	1.2
Australia	2.9	1.9
Others	30.6	20.3
Total	151.0	·1.0
North American countries	6.3	4.2
European Economic Community[a]	66.9	44.3
COMECON countries[b]	20.2	13.4
Others	57.6	38.1

Source: International Iron and Steel Institute, *World Steel in Figures, 1980* (Brussels: IISI, 1980), p. 16.
a. Includes Belgium, Denmark, France, West Germany, Ireland, Italy, Luxembourg, Netherlands, and United Kingdom.
b. COMECON (Council for Mutual Economic Assistance) countries include Bulgaria, Czechoslovakia, East Germany, Hungary, Poland, Romania, and the Soviet Union.

ductivity increased dramatically. Moreover, the Japanese led the way in applying sophisticated computer control to the pouring, forming, and rolling of steel products. This increased the yield of finished products from raw steel and thereby reduced man-hours required per net finished ton.

Japanese wages increased by 244 percent (in U.S. dollars) in the decade following the 1958 recession, but unit labor costs actually declined by more than 30 percent. During the same period U.S. wage rates rose by only 39 percent, and unit labor costs remained constant.[6] Thus in the decade after the 1958 recession Japanese material costs, unit labor costs, and shipping costs fell substantially, while U.S. labor and material costs remained virtually constant, but U.S. surface transportation costs rose. These developments could only place the U.S. industry under import pressure unless protective governmental policies were erected.

Whatever the cause, the U.S. industry grew much more slowly after 1959.

6. Japanese and U.S. wage data are from Federal Trade Commission, *The United States Steel Industry and Its International Rivals: Trends and Factors Determining International Competitiveness* (GPO, 1978), tables 3.1 and 3.3, pp. 113 and 117.

Between 1956 and 1959 it had increased its raw steel capacity by 10 million tons, from 130 million to 140 million tons.[7] By 1970 raw steel capacity was only 154 million tons (table 2-4). Thus the rate of expansion slowed from 2.5 percent a year in the late 1950s to only 1 percent a year in the 1960s.

In 1956 the industry produced 115 million tons of raw steel. For the first half of the decade of the 1960s, output fell to an average of 106 million tons a year. Part of the decline was due to imports, which averaged 4.5 million net tons of mill products (equivalent to approximately 6.3 million raw tons at U.S. yields), but probably more important was the very low level of steel consumption. With the fiscal stimulus of 1964 and the ensuing Vietnam War, steel consumption rose sharply, from an average of 108 million net tons of crude steel a year in 1960–64 to an average of 146 million tons in 1965–69. Raw steel production, on the other hand, rose only to an average of 133 million tons a year in the last half of the decade as imports continued to accelerate. Hence, even in the late 1960s the industry was unable to reach a production level that could have placed pressure on its 1959 capacity (140 million tons). Profits remained low as import competition grew increasingly acute, and the industry would add less than 5 million tons of new capacity in the 1970s.

The Industry's Financial Performance

Given the dramatic change in the fortunes of the U.S. industry, it is not surprising that profit rates plummeted in the 1960s. From 1950 through 1957 the industry generally earned a slightly higher rate of return than the average U.S. manufacturing industry. For the next twenty years the industry's average return on equity would be approximately 25 percent below the manufacturing average (table 2-7).

Was the recession of 1958 initially viewed by the industry and investors alike as only a temporary setback for the industry? With the recovery of economic activity, investors might have expected steel companies to recover to their respectable performance of the early 1950s. Surprisingly, the stock market turned bearish on the industry in 1958 and began to place downward pressure on steel equities—pressure that would not abate until the commodities boom of 1973–74. The ratio of market to book value of steel equities

7. All data in this paragraph are based on estimates of capacity derived from peak production periods (see table 2-4).

Table 2-7. *Rate of Return on Equity after Taxes: Steel versus All Manufacturing, 1950–78*[a]
Percent

Year	*Primary U.S. iron and steel firms*	*All U.S. manufacturing*
1950	14.3	15.4
1951	12.3	12.1
1952	8.5	10.3
1953	10.7	10.5
1954	8.1	9.9
1955	13.5	12.6
1956	12.7	12.2
1957	11.4	11.0
1958	7.2	8.6
1959	8.0	10.4
1960	7.2	9.2
1961	6.2	8.8
1962	5.5	9.8
1963	7.0	10.3
1964	8.8	11.6
1965	9.8	13.0
1966	10.3	13.5
1967	7.7	11.7
1968	7.6	12.1
1969	7.6	11.5
1970	4.3	9.3
1971	4.5	9.7
1972	6.0	10.6
1973	9.5	12.8
1974	16.9	14.9
1975	10.9	11.6
1976	9.0	14.0
1977	3.6	14.2
1978	8.9	15.0

Source: Based on data in U.S. Federal Trade Commission-Securities and Exchange Commission, *Quarterly Financial Reports for Manufacturing Corporations* (Government Printing Office, 1950) and subsequent quarterly issues through 1978.
a. After-tax rate of return on book value of stockholders' equity.

plummeted from 1.6 to less than 0.6 in twelve years (figure 2-1).[8] In contrast, the general stock market continued to rise until the mid-1960s, but the average industrial issue reached a plateau in the late 1960s and began to slide in the 1970s. Thus the market began to discount bad news in the steel industry long before the general weakness in industrials developed. Apparently investors

8. This analysis is based on Standard and Poor's indexes for steel and for the 400 S&P Industrials.

Figure 2-1. *Ratio of Market to Book Value of Equity, 1946–78, Steel Industry versus Standard and Poor's 400 Industrials*[a]

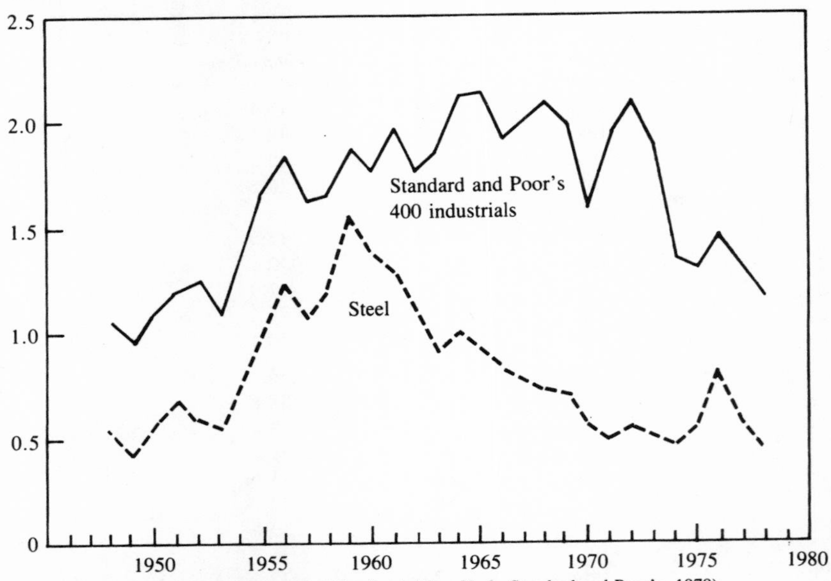

Source: Standard and Poor's, *Analysts Handbook* (New York: Standard and Poor's, 1979).
a. This ratio is the average of high and low market values for the year divided by the book value of equity on December 31.

recognized the fundamental difficulties ahead as early as 1958–59, long before the confrontation between President Kennedy and the industry in 1962. Imports may have served as the early warning signal.

Despite the sharp fall from prosperity, the steel industry has not become more risky in the eyes of investors. The measure of systematic risk for steel equities—the beta coefficient—has remained very close to unity for decades, reflecting an average degree of risk for investors.[9] Aluminum, motion pictures, forest products, and chemicals are much more risky than steel. Thus investors apparently continue to see little excess risk in holding steel equities. Even though steel equities are now priced at 40 percent of their historical book value, investors apparently believe there is no major risk of further deterioration of the industry's position. Holding steel equities carries about the same risk as holding the average stock on the New York Stock Exchange.

9. See Gerald A. Pogue, "Estimation of the Cost of Capital for Major United States Industries with Application to Pollution-Control Investments" (U.S. Environmental Protection Agency, November 1975), chap. 4.

Price Behavior since the 1950s

The steel industry has often been characterized as an oligopoly practicing "administered pricing." Adelman's theory of price adjustment in the 1950s suggests that the industry was continually below cartel equilibrium until 1959 and was using wage negotiations as signals for price increases.[10] In this situation temporary reductions in demand would not be expected to lead to price cutting. After 1959, however, import competition might have made prices more responsive to demand pressures.

In fact, most empirical studies of steel pricing find little cyclicality. In their analysis of buyers' prices and the wholesale price index, Stigler and Kindahl found that steel prices tended to deviate very little from the list prices recorded in the index in the period 1957–66.

> Neither index displays a noticeable cyclical movement in either expansion or contraction. Nor are the short-run fluctuations of appreciable size.
>
> This finding, it must be confessed, comes as a surprise to us. The steel industry is now unconcentrated as compared with the first decade of the century, or indeed as compared with many other industries in our sample. Import competition was growing fairly steadily during the period. With the exception of three steel products, however, we were not able to learn of any important and continuous departures from quoted prices. The exceptions were reinforcing bars . . . , pipe, and stainless steel products.[11]

The exceptions cited by Stigler and Kindahl are all products produced by small firms as well as by the major integrated producers. Thus for the period of their study, there was no evidence of price cutting for the products produced by large, integrated firms alone—flat-rolled and heavy bar products.

Similar results have been found by Jondrow, who was unable to detect an effect of changes in capacity utilization on list prices.[12] Mancke, on the other hand, finds a significant effect of capacity utilization, particularly after 1958, when the rites of spring were suspended.[13] Rippe's results are more ambiguous since he finds a decided break in the structure of his wholesale price-change equation in 1959.[14] Hersh reports that she is able to find a small positive

10. Adelman, "Steel, Administered Prices, and Inflation," pp. 16–40.

11. George J. Stigler and James K. Kindahl, *The Behavior of Industrial Prices* (National Bureau of Economic Research, 1970), pp. 73–74.

12. James M. Jondrow, "Effects of Trade Restrictions on Imports of Steel," in U.S. Department of Labor, Bureau of International Labor Affairs, *The Impact of International Trade and Investment on Employment* (GPO, 1978), pp. 11–25.

13. Mancke, "The American Iron Ore and Steel Industries," p. 197.

14. Richard D. Rippe, "Wages, Prices, and Imports in the American Steel Industry," *Review of Economics and Statistics*, vol. 52 (February 1970), pp. 34–46.

effect of capacity utilization on list price, but only when the industry's capacity utilization rises above 80 percent.[15] When utilization falls below this level, further price cutting does not break out; at least list prices are not reduced. She also reports, however, that list and realized prices diverge widely at times, but she does not attempt to estimate a realized price equation.

A report by the Council on Wage and Price Stability compares realized price data (Census Bureau) with Bureau of Labor Statistics price data from the wholesale price index (now producers' price index).[16] The gap between the wholesale price index list price and the transactions price (Census Bureau unit values) has been widening during successive recessions since 1958. The results are not uniform across all products nor all post-1958 recessions, but there does appear to be a growing cyclicality in pricing in the industry. As the next chapter will show, much of this has been driven by foreign steel prices, which in turn are sensitive to world demand conditions.

In 1979 the Antitrust Division of the Department of Justice acknowledged that it was investigating the industry for possible violations of the Sherman Anti-Trust Act during 1976.[17] The investigation apparently involved agreements on the form and size of "extras," costs added to base prices for the myriad sizes, shapes, and grades of steel. No charges were ever brought, but the evidence would suggest that prices are becoming more flexible with time.

Employment and the Problem of Labor Intensity

One cannot assess the steel "problem" without discussing the ramifications for labor and labor's role in the decline. Obviously, in the absence of industry growth, employment has been falling for several decades. Advances in technology and some deepening of capital have led to a lower labor-capital ratio. Without offsetting increases in demand for U.S.-produced steel, the steel labor force has shrunk.

Raw steel production recovered from its post-1958 recession doldrums in the mid-1960s, averaging more than 133 million tons from 1965 through 1969. By 1978, a relatively good year, it was only 137 million tons. During

15. Rosanne M. Hersh, "Industry Economic Service: The Steel Industry" (Goldman, Sachs, 1977), p. 29.

16. U.S. Council on Wage and Price Stability, *Report to the President on Prices and Costs in the United States Steel Industry* (COWPS, October 1977), app. table 15, p. 119. All realized price data are derived from U.S. Bureau of the Census, *Current Industrial Reports: Steel Mill Products,* Series MA-33B (GPO), selected issues. They are calculated by dividing the total value of shipments by total tonnage; therefore, they are not fixed-weight price indexes.

17. *American Metal Market,* June 13, 1979, p. 1.

this period total industry employment fell from slightly more than 650,000 to slightly more than 550,000 employees.[18] Production workers totaled 538,000 in 1965 but only 437,000 in 1978. Productivity grew by 1.4 percent a year from 1966 through 1978, while total output grew by only 0.2 percent a year.[19]

With the modest productivity growth since the 1960s, the industry remains rather labor-intensive. The ratio of labor payments to total value added has been extremely high in this industry even during prosperous years such as 1974. Between 1972 and 1976 manufacturing industries in the United States paid an average of 51 percent of value added to workers in the form of wages and supplementary benefits. For the same period the steel industry paid labor an average of 64 percent of its value added, even though 1973 and 1974 were boom years for steel. The steel industry by this measure is even more labor-intensive than the apparel or electronics industries, which averaged 58 and 57 percent, respectively.[20] It can hardly be argued that these latter two industries are ones in which the United States enjoys a comparative advantage.

Why had the U.S. steel industry not substituted capital for labor more rapidly? As will be shown, it is not because low wage rates dictate a high labor-capital ratio. Barnett argues that the industry has not moved more aggressively in this fashion since the 1950s because it could not afford to build modern, new plants.[21] With low rates of return on capital, the industry was left with no choice other than to plow its limited cash flow back into tired, old plants. But this suggests that new plants could or should have been built and that these new plants would have generated major improvements in productivity, a judgment examined critically in chapters 4 and 5.

How much substitution is possible? Barnett demonstrates that the U.S. industry currently (in 1978) uses 6.7 man-hours per net raw ton of steel produced. This translates into 9.4 man-hours per net ton of finished products shipped and reflects the total mix of alloy, stainless, and carbon products. For the carbon steel sector, the average labor content per ton of finished shipments is probably slightly more than 8.0 man-hours at full capacity. The

18. U.S. Department of Labor, Bureau of Labor Statistics, *Employment and Earnings*, selected issues, 1958–78.

19. U.S. Department of Labor, Bureau of Labor Statistics, *Productivity Indexes for Selected Industries, 1979* (GPO, 1979), p.123.

20. Data in this paragraph are from U.S. Bureau of the Census, *Annual Survey of Manufactures: Industry Profiles, 1976*, Series M76(AS)-7 (GPO, 1978), pp. 70, 163.

21. Donald Barnett, "Labor Productivity and Capital Formation in the U.S. Steel Industry," paper presented at the George Washington University Conference, "The American Steel Industry in the 1980's—The Crucial Decade," Washington, D.C., April 19, 1979.

Japanese industry, most of which has been built in the past fifteen years, requires about 6 to 6.4 man-hours per net raw ton, but its yield to finished products is somewhat greater, probably about 0.83 for a comparable product mix.[22] Therefore, the Japanese industry—the paragon of efficiency—requires 7.5 man-hours per finished ton at 85 to 90 percent capacity utilization. For carbon steel products alone, this number is between 7.0 and 7.5 man-hours per ton for 1978, assuming 85 to 90 percent capacity utilization.

As chapter 4 demonstrates, there is little variance in the labor-capital ratio across various new plants. The Japanese employment costs per hour of labor in the steel industry in 1978 were approximately $8.22, while the U.S. costs were $14.04. This difference widened in 1979–80 to nearly $7 an hour as the yen depreciated.[23] Despite this difference the optimal new plant in Japan and the United States would look remarkably similar according to steel engineers. Therefore, if the U.S. industry had the opportunity to build new mills at the same rate as the Japanese over the past two decades, it might have lowered its labor usage to the Japanese level, or by about 10 percent. This could have reduced unit labor costs in 1978 by at most $12 per raw ton for carbon steel products, a reduction that in itself could hardly restore the industry to its position of the 1950s.

The Increase in Labor Rates

Given the labor intensity of steel production and the limited possibilities for substituting capital for labor in new plants, the cost of labor plays a key role in determining comparative advantage. In labor-scarce economies steel is likely to look like a bad investment. A single new 6-million-raw-ton plant with a full array of finishing facilities will require a labor force of more than 12,000 workers. Many of the Japanese plants are one and one-half to two times this size; hence they require truly mammoth work forces.

When the U.S. industry had the unique advantages of an inland location, proximity to cheap raw materials, and a growing industrial complex with a voracious appetite for steel, rising labor costs might not have posed much of a problem. But when shipping costs fell, world iron ore and coal prices fell in real terms, and developing countries with very low wage rates began to

22. See "Treasury Announces Task Force Revision, Fourth Quarter Adjustment, Product Changes, and Great Lakes Freight Rate Adjustments," U.S. Department of the Treasury press release, July 20, 1978.

23. Peter F. Marcus and others, *World Steel Dynamics: The Steel Strategist* (Paine Webber Mitchell Hutchins, Inc., October 1980).

build major steel industries, the level of the U.S. wage began to matter very much.

As table 2-1 demonstrated, labor costs in the steel industry rose very rapidly in the early 1950s. The rate of increase began to slow in the 1960s in response to weak demand and the aftermath of the 1959 strike. Between 1957 and 1967 hourly employment costs in the steel industry rose at about the same rate as in the average manufacturing industry (see table 2-8). Thereafter, with the 1968 settlement, steel wages began to rise sharply once more. For example, in 1967 average compensation in steel was approximately 38 percent above the manufacturing average; by 1974 this had risen to 60 percent, and by 1977 steel compensation was 71 percent more than the average compensation for all manufacturing. The 1974 and 1977 settlements were major contributors to this escalation.

It is interesting to examine relative wage patterns in manufacturing across developed economies. The premiums enjoyed by auto workers and steelworkers are much larger in the United States than in Japan or Germany, for example. As figure 2-2 demonstrates, these premiums are much more closely bunched in the electronics, paper, and machinery industries. The U.S. relative wage is much the highest in steel and autos but not in other industries. Given the large share of labor payments in value added in these industries, it is hardly surprising that the U.S. automobile and steel industries have major "import problems."

In 1973 the industry gave large concessions to the United Steelworkers in return for an Experimental Negotiating Agreement (ENA), which limited the union's right to strike during future contract negotiations in return for a wage increase of 3 percent a year for the next three years, a liberalized cost-of-living escalator, and a $150 bonus.[24] Thus the industry locked itself into providing 3 percent wage increases each year plus an escalator that reimbursed workers for two-thirds to three-quarters of the recorded rate of inflation. With no productivity growth after 1973, this agreement could only be described as expensive. More important, the ENA provided a floor under the 1974 and 1977 negotiations that the industry could not violate without destroying the no-strike basis of the agreement.

Why did the steel industry allow its labor rates to escalate so rapidly in the 1970s, after succeeding in moderating wage increases in the 1960s? There are many reasons.

First, in 1971 and then in 1973 the U.S. dollar was devalued. This provided steel producers with some breathing room after years of steadily rising imports.

24. *New York Times*, March 30, 1973.

Table 2-8. *Hourly Labor Costs in the U.S. Steel Industry and in All Manufacturing, 1956–78*

Year	*Hourly earnings*[a] *(dollars)* Steel industry (BLS)	Steel industry (AISI)	Manufacturing (BLS)	*Total compensation for all employees (dollars)* Manufacturing (BLS)	Steel[b] (AISI)	*Value of the dollar (trade-weighted index, March 1973 = 100)*
1956	2.57	2.54	1.95	2.40	3.15	115.27
1957	2.73	2.73	2.04	2.54	3.45	117.90
1958	2.91	2.93	2.10	2.65	3.79	117.82
1959	3.10	3.14	2.19	2.76	4.07	119.89
1960	3.08	3.09	2.26	2.87	4.08	119.90
1961	3.20	3.24	2.32	2.95	4.27	119.13
1962	3.29	3.33	2.39	3.06	4.44	119.45
1963	3.36	3.39	2.45	3.16	4.52	119.60
1964	3.41	3.43	2.53	3.29	4.61	119.60
1965	3.46	3.54	2.61	3.35	4.71	119.67
1966	3.58	3.64	2.71	3.50	4.90	119.86
1967	3.62	3.66	2.82	3.68	5.06	119.95
1968	3.82	3.86	3.01	3.94	5.35	122.08
1969	4.09	4.12	3.19	4.20	5.71	122.38
1970	4.22	4.24	3.35	4.50	6.05	121.08
1971	4.57	4.57	3.57	4.78	6.68	117.75
1972	5.17	5.22	3.82	5.03	7.47	109.06
1973	5.61	5.69	4.09	5.39	8.10	98.85
1974	6.41	6.55	4.42	5.95	9.54	101.30
1975	7.12	7.23	4.83	6.66	11.10	98.32
1976	7.79	8.00	5.22	7.22	12.22	105.53
1977	8.59	8.91	5.68	7.83	13.42	103.26
1978	9.70	9.98	6.17	8.47	14.69	92.16

Source: Selected issues of the *Annual Statistical Report: American Iron and Steel Institute*, which contain the BLS data as well as the AISI data.
a. BLS data exclude office, clerical, and supervisory personnel.
b. Nonsalaried workers.
BLS = Bureau of Labor Statistics.

Figure 2-2. *Ratio of Industry Wage to Average Manufacturing Wage in Selected Industries in the United States, Germany, and Japan, 1978*

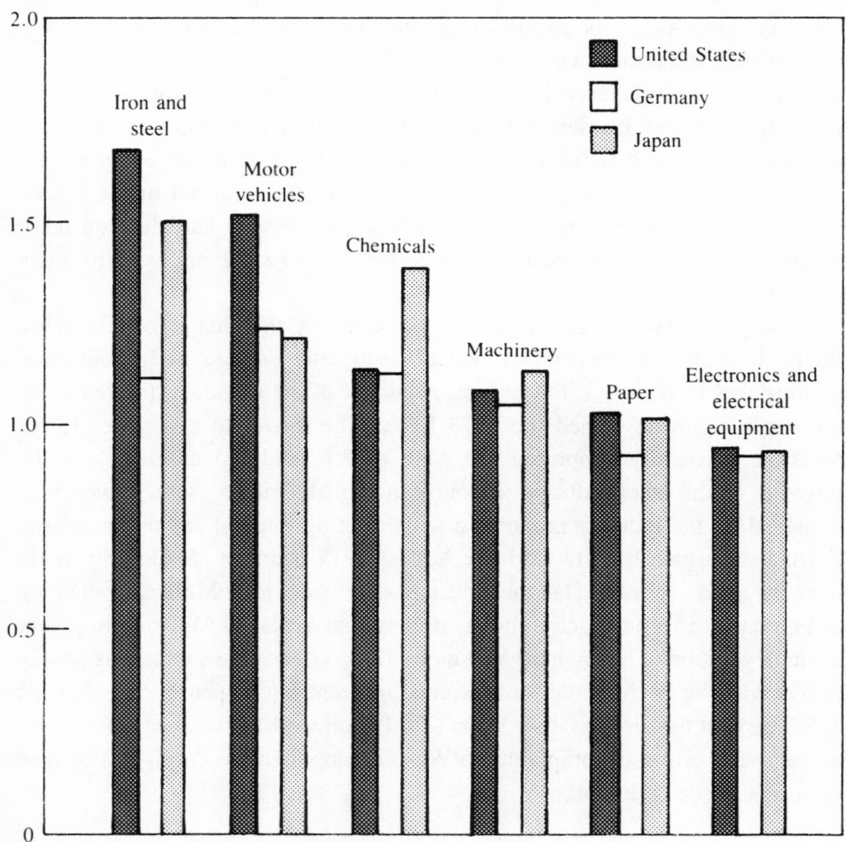

Source: U.S. Department of Labor, Bureau of Labor Statistics, unpublished data for 1978.

The years 1972–73 were characterized by rising production, increasing exports, and declining imports. The industry was becoming more profitable, and many steel executives felt they could buy a permanently larger share of the market through domestic labor peace.

Second, the 1969–74 Voluntary Restraint Agreements on imports were in full flower in 1971–72. With a falling value of the dollar, reduced imports from Europe and Japan might have been expected anyway, but the voluntary quotas undoubtedly reduced the competitive price pressure from abroad. The result was steadily rising profit margins.

Third, the timing of labor negotiations was especially cruel to the industry. Its 1974 contract expired in the middle of a price explosion in the world export market. The 1973–74 period saw the U.S. industry operating at almost full capacity, increasing its exports, and generally sharing the benefits of a worldwide commodity boom (table 2-4). In 1973 the industry wanted to be sure it was not foreclosed from sharing in this bonanza by fears of a strike in 1974. Therefore, it essentially settled with the United Steelworkers eighteen months before the expiration of the contract in 1974, at a very steep price.

Fourth, 1974 was the year in which price controls expired in the United States. One could not expect an industry to take a very hard line on labor negotiations at a time of general excess demand for steel and expiring price controls.

Finally, the 1977 negotiations were clouded by the change of leadership in the United Steelworkers. A radical insurgent, Edward Sadlowski, was arguing that I. W. Abel, the retiring president of the union, had given away too much when he signed the 1973 ENA. The moderate candidate, Lloyd McBride, generally supported the Abel policy, and it was clearly in the interests of the steel industry to contribute to McBride's victory over Sadlowski. Had the industry mounted a winter-spring internal campaign to drop ENA and bargain hard toward the August 1977 deadline, Sadlowski might have defeated McBride. Had the industry waited until after McBride's election in February to launch such a thrust, it would have placed McBride in a very difficult position in his own organization. The conservative course was simply to live with the ENA, settle for an annual increase in compensation estimated at 9.3 percent for the next three years (if inflation were to continue at 6 percent a year), and bring its complaints to Washington when the floodgates opened on a new wave of imports.[25]

The Burden of Regulation

Steel production involves a large number of dirty, dangerous processes. The generation of air- and water-borne residuals has historically been one of the unfortunate side effects of steelmaking, and many of these residuals are potentially very harmful to workers and to nearby residents. In addition the huge equipment, handling very hot and extremely heavy products, creates major worker-safety problems. Even in an unregulated world, steel companies

25. Council on Wage and Price Stability, *Report to the President on Prices and Costs in the United States Steel Industry*, pp. 32–34.

would be forced to expend resources on worker protection. They might not, however, be very concerned about residuals generation into the general environment unless the capture of these residuals yielded considerable revenues from the sale of by-products.

The cost of environmental regulation in the steel industry is not easily measured because of the difficulty in allocating operating and capital costs, the imprecision of the regulations affecting the industry, and the rather uneven enforcement pattern across plants. The Environmental Protection Agency (EPA) commissioned a study of the costs of environmental controls in this industry that provided hypothetical estimates of the cost of full compliance with regulations pending before 1977. This study concluded that by 1983 the average cost of steel products would be increased by 4.6 percent by the pollution control outlays required in the 1975–83 period, or $16.75 a net ton in 1975 dollars.[26] To this must be added $3.25 in costs associated with pre-1975 investments. Thus the ultimate 1983 cost of environmental regulations is $20 a ton in 1975 dollars, or 5.5 percent of 1975 prices. In 1978 prices this would add $21 to the price of the average carbon steel product. Slightly more than half of these price effects would be registered by 1977, according to the study, assuming full compliance with applicable regulations, but the industry was far from full compliance in 1977, a situation that continued into 1981. Therefore, the cost of environmental regulation was undoubtedly less than $10 per ton of output in 1978–79.

Another approach to estimating the current cost of environmental regulations in the industry is to use reported outlays on environmental control equipment. In 1976–78 the industry spent an average of $494 million a year, up sharply from earlier years.[27] If 20 percent is added to this figure for operating costs, the average annual expenditure in current dollars for pollution control would still be less than $600 million a year. Spread over annual shipments of 93 million tons, these outlays average only $6.37 a net ton. Environmental outlays will undoubtedly rise above this level, and the full capital and operating costs may eventually rise to the estimated 5.5 percent of product prices (without controls), but the costs could hardly be said to be a major cause of the industry's current problems.

Occupational health and safety programs also add to the industry's regu-

26. Temple, Barker, and Sloane, Inc., "Analysis of Economic Effects of Environmental Regulations on the Integrated Iron and Steel Industry" (U.S. Environmental Protection Agency, July 1977), pp. 6-6 and 6-7.

27. *Annual Statistical Report: American Iron and Steel Institute, 1978* (Washington, D.C.: AISI, 1979), p.10.

latory burdens, but by the industry's own estimates these costs are only about one-sixth of environmental costs.[28] If this ratio holds through 1983, the full costs of occupational health and safety and environmental policy might be as much as 6.4 percent of the price of steel without controls. At present, however, there is reason to believe that the industry's cost is less than half of this ultimate potential.

The conclusions above should not be taken as evidence that regulatory costs should be ignored. Annual costs of $700 million a year are obviously not trifling, nor is it clear that these costs could not be reduced substantially without sacrificing worker health or environmental quality. A shift away from engineering standards to performance standards or economic incentive systems might permit considerable savings in control costs. The EPA has already made a modest start in this direction by allowing firms to choose control techniques (with some limitations) that achieve plant-wide standards for air pollution. Such regulatory innovations could reduce the industry's regulatory burden, but they will not be sufficient to return the industry to a position of economic health.

Trade Policy and the Crisis of 1977

The declining fortunes of the steel industry were countered with surprisingly little protectionism for ten years, perhaps because of the image the industry projected in confrontations with the government over wage and price policy that began in the Truman administration. During most of the Eisenhower administration, because of a generally low inflation rate and an apparently healthy industry, little conflict arose. The industry's problems began in earnest during the last two years of Eisenhower's last term, but it was not until 1962 that the next collision between Big Steel and government was to occur.

The confrontation between President Kennedy and Roger M. Blough, chairman of United States Steel, took place in 1962 over the industry's price response to the 1962 labor settlement. Eventually U.S. Steel and its rivals, led by Inland, rescinded their 1962 list price increases, but foreign competition might have forced this result anyway. Steel prices in the United States were essentially flat from 1959 through 1964 for most products and through 1968 or 1969 for some. With the increase in demand occasioned by the Vietnam War boom, prices began to rise once again in the late 1960s. In 1968, during

28. American Iron and Steel Institute, *Steel at the Crossroads: The American Steel Industry in the 1980's* (AISI, January 1980), p. 43.

another year of wage negotiations, President Johnson once again applied jawboning pressure to the industry, eventually succeeding in getting a rollback in list prices.[29]

Perhaps as a result of these confrontations over "administered pricing" of steel, the industry was unable to press for increased trade protection in the 1960s even though imports rose dramatically. It was not until 1968 that the industry was able to muster sufficient political muscle to attempt to obtain relief from imports in a major way. In that year the State Department negotiated a set of Voluntary Restraint Agreements with European and Japanese exporters.[30] The effect of these agreements is examined in chapter 5, but casual evidence suggests that the impact could not have been very great. Certainly the resurgence of imports in 1971, a labor negotiation year, to 18.3 million net tons suggests that the reduction in imports due to the VRAs could not have been large. The years 1973 and 1974 were years of extraordinarily strong world steel demand. Imports were not a problem in these years nor immediately in the recession that followed.

In 1977, however, just as the world economy appeared to be recovering well from the recession, the growth in world steel demand slowed. World raw steel production fell in 1977 by a very small amount after growing by a modest 5 percent in 1976 (table 2-4). The result was that European steel producers began to compete vigorously for market share in export markets and within the EEC. Prices fell substantially in Europe and in the U.S. import market in late 1976 and early 1977, and steel imports into the United States began to increase in mid-1977. The downward pressure on U.S. producer prices (as U.S. mills moved to match some of the price reductions) and the reduced volume of steel output led to severe financial pressures on some U.S. steel producers. The industry's profit margin fell to less than 1 percent on sales. One company, Alan Wood Steel, went bankrupt and closed its only plant in 1977. Youngstown Sheet and Tube and Bethlehem Steel announced the partial closing of three separate plants. Youngstown closed most of its Campbell Works in Youngstown, Ohio, while Bethlehem idled steelmaking capacity in Lackawanna, New York, and Johnstown, Pennsylvania. The resulting layoffs placed enormous pressure on Congress and the Carter administration to do something to prevent what politicians feared was a further collapse of the industry.

29. For a discussion of this confrontation, see Federal Trade Commission, *The United States Steel Industry and Its International Rivals*, pp. 267–80.

30. See Wendy E. Takacs, "Quantitative Restrictions on International Trade" (Ph.D. dissertation, Johns Hopkins University, 1975), for a discussion of the VRAs.

In fact, there was little likelihood that many more plants would close in the wake of the import bubble of 1977,[31] but a Steel Caucus quickly formed in Congress, enlisting more than 200 members at its peak. In the late summer of 1977 the Carter administration was forced into taking some action to head off the political pressures. In a meeting with steel executives, organized labor, academics, and legislators, President Carter invited the industry to pursue relief through the newly enacted amendments to the Antidumping Law contained in the Trade Act of 1974.[32] The definition of dumping had been extended to sales below cost even if export prices were no lower than domestic prices. In essence the president was responding to charges of dumping against European and Japanese steel producers by inviting U.S. firms to file charges with the Treasury Department. At the same time, he asked the Council on Wage and Price Stability to prepare a report on the industry's condition.

The results President Carter obtained on his two initiatives were somewhat contradictory. On the one hand steel companies filed dumping complaints, alleging dumping by Japanese and EEC producers. On the other hand, while the council's report demonstrated that Japanese firms indeed enjoyed a substantial cost advantage over U.S. producers, it suggested strongly that dumping complaints against European exporters, based on the new cost-of-production test in the Antidumping Law, would probably succeed. When National Steel and Armco filed dumping cases against the European exporters, the administration faced a major dilemma. If it simply processed the dumping complaints against European and Japanese exporters, it would sharply curtail European exports to the United States. The European industry was operating with substantial excess capacity, and it could not hope to cover unit costs in sales to the United States, as apparently required by the 1974 amendments to the Antidumping Law. Japanese exports could easily supplant European products in the U.S. market, since Japan also had nearly 30 percent excess capacity, but the act of denying European exporters access to the U.S. market would poison U.S.-European trade relations. Such an action would surely destroy any chance of consummating the multilateral trade negotiations then in progress and would fan the flames of inflation.

In an effort to steer a middle course Carter asked Under Secretary of the

31. A few more plants have closed since 1977. (See chapter 8.) The raw steel capacity shutdown in the ensuing three years, however, has been relatively modest.

32. A description of the new provisions of the Antidumping Act is found in "Report to the President: A Comprehensive Program for the Steel Industry" (December 6, 1977), pp. 10–13. This document was known as the Solomon Report because Anthony Solomon, an under secretary of the Treasury, was chairman of the task force studying steel industry issues. Hereinafter referred to as the Solomon Report.

Treasury Anthony Solomon to develop a policy in just a few weeks. The Solomon Task Force began meeting in September and finished its work on December 1, submitting a set of recommendations to the president which he accepted in full.[33] Solomon, who had negotiated the VRAs in 1968 during his tenure at the State Department, suggested a "reference price" system that would serve as a basis for monitoring imports to determine if there were prima facie evidence of dumping. These reference, or "trigger," prices, as they were eventually called, were based on estimated Japanese costs of production at standard volume. This device allowed Solomon to keep European steel flowing to the United States at "fair" prices and to finesse the very troubling cost-of-production standard in the 1974 Trade Act. The Japanese cost of production would be calculated at a five-year average rate of capacity utilization—85 percent—rather than the 72 percent at which the industry operated in 1977. In this manner import protection could be afforded U.S. producers without disastrous domestic price effects.

The Solomon Plan was welcomed by U.S. producers and exporters since it was seen as a device for propping up world prices without lengthy legal proceedings. At the same time, Commissioner Etienne Davignon of the European Commission was devising his own reference price system for Europe. Prices in world trade rose substantially in 1978, partly because of a 6 percent growth in demand, but also because of the U.S. and European protectionism. The Japanese, initially wary, were surprised to find that these policies plus an appreciating yen allowed them to operate their five largest producers at a profit, at an operating rate of little more than 70 percent.

The U.S. government suspended the trigger price system in March 1980, when U.S. Steel filed a dumping complaint against the producers in five European countries. This was in keeping with Solomon's assertion that the government could not enforce the trigger price system and prosecute dumping complaints.Throughout the 1978–80 period, however, the Japanese were limiting their exports to the United States to about 6 million tons a year per an apparently informal agreement with U.S. trade officials. As a consequence, suspension of the trigger price system did not lead to increases in Japanese exports to the United States.

In late 1980 the Europeans took a major step toward cartelizing their steel industry. Davignon admitted failure in attempting to keep internal steel prices high through minimum-price regulation. Fearing that competition among producers was about to break out in earnest, he succeeded in persuading the European Commission to declare the industry to be in a situation of "manifest

33. Ibid.

crisis'' and to impose production quotas on the EEC's steel producers.[34] The Europeans had clearly indicated a willingness to go much further than the United States to shore up their overbuilt, inefficient industry.

The effects of the Solomon trigger prices on U.S. prices and import shares will be examined in chapter 5. For the present it is important to note that the trigger prices were part of the first comprehensive plan of governmental assistance to the steel industry, an industry that had been in trouble since the 1958 recession (with the exception of the 1973–74 commodities boom period). Solomon also proposed extending loan guarantees to firms with limited access to capital markets and plants in declining steel-producing regions, a reexamination of EPA policy toward the steel industry, an examination of the possibility of increasing the allowed rate of depreciation of steel assets for tax purposes, and the creation of a tripartite committee to develop further ideas. The key to the program was the trigger price system, but the attempt at a comprehensive approach prompted Japanese officials to remark optimistically that this might be the beginning of a U.S. industrial policy similar to that carried out by the Ministry of International Trade and Industry in Japan.[35] This was not the intention of the Solomon group, but the program marked a major turning point nevertheless. Steel was no longer to be the government's scapegoat for inflation; in fact, it was to be nurtured and protected from unfair foreign competition. In October 1980 the trigger price system was reinstituted when U.S. Steel suspended its dumping charges against the European producers. How a protective, rather than a hostile, government moves from this makeshift policy to longer-term policies will be critical to the development of the domestic steel industry in the next few decades.

Summary

The U.S. steel industry has found it difficult to compete with imports of steel from Japan and the less developed world. Much of this difficulty originated in the late 1950s, when iron ore deposits were being discovered in various parts of the world, shipping rates were falling, and U.S. steel wages were rising sharply. New exporters, such as Japan, began to flourish in the 1960s, and the United States became their best market. As a result,

34. ''Commission Decision . . . of 31 October 1980 Establishing a System of Steel Production Quotas for Undertakings in the Iron and Steel Industry,'' *Official Journal of the European Communities,* vol. 23, L291 (October 31, 1980), p. 1.

35. Conversation between the author and MITI officials, May 1978.

U.S. capacity did not grow, and existing capacity was underutilized in all but three years from 1958 through 1977. Only two new integrated plants have been constructed in the United States since World War II. Capacity growth has been limited to increments obtained through rounding out or replacing older facilities and the construction of mini-mills. Imports have expanded to nearly 20 percent of the U.S. market.

In 1968 and again in 1977 the industry was able to obtain a measure of relief from growing import pressures through ad hoc governmental policies. The 1969–74 Voluntary Restraint Agreements were a set of voluntary quotas agreed to by European and Japanese exporters. The 1977 measures were far more comprehensive, embodying a set of reference, or trigger, prices against which import prices are to be measured. These trigger prices became the centerpiece of the Solomon Plan, which was designed to rescue the industry from its doldrums in late 1977. Whether this plan has worked and whether it should be continued are issues to be examined in this monograph. First, however, a model of world and U.S. price behavior and U.S. market share will be developed before an attempt is made to measure the effects of trade protection and to examine the rationale for such a policy.

III

A Model of Pricing and Market Shares

This chapter presents a detailed model for explaining U.S. import prices, U.S. domestic prices, and the import share of steel in the U.S. market. The model is estimated for a set of steel mill products: hot-rolled sheet, cold-rolled sheet, plate, hot-rolled bars, and structurals. This ameliorates the problems created by shifts in product mix over time and thus reduces the errors in variables problems.

The purposes of the model are to demonstrate how competitive influences have affected U.S. producers' prices in the past two decades, to determine if export prices reflect competitive influences, and to provide an explanation for the timing and magnitude of the loss of market share by U.S. producers. Specifically I wish to determine if U.S. producers' prices respond to world market conditions as well as to U.S. market conditions. To what extent are prices inflexible in the face of import pressure? Moreover, given the repeated assertions about the motives of foreign producers, particularly the Japanese, it is important to determine if export prices from the leading exporter, Japan, reflect world market conditions or simply a desire to maintain employment levels. Are the Japanese exporting unemployment or are they simply responding to changes in world demand and to their own cost pressures?

The Setting

As demonstrated in chapter 2, the world steel industry underwent a major change in the late 1950s. Recovery from World War II, the development of new iron ore deposits in various parts of the world, and declining shipping costs during the first two postwar decades led to the growth of numerous new steel producers and a much more competitive world market. One would therefore expect prices of steel in export markets to reflect competitive forces for the period since the 1958 recession.

Obviously not all potential import markets for steel may be characterized as part of a single, competitive whole. Steel is an important industrial product, and many governments (including the United States) are likely to want to control imports, stimulate exports, or both. Thus it should not be surprising if prices for steel in various markets differ by more than the cost of transportation among them. But after 1958 the U.S. market rather quickly became the world's largest export market, and until 1969 U.S. policy generally allowed relatively free importation of foreign steel to compete with U.S. products after the payment of modest tariffs.[1] With a large number of foreign suppliers competing for this business, one would expect import prices to reflect changes in exporters' factor prices, shipping rates, and world demand if these suppliers compete actively.

Since there are numerous potential and actual sources of supply for each imported steel mill product, one cannot measure costs in each country and assume that these costs represent the opportunity cost of resources used in supplying a ton of steel to the United States. Instead, I assume that Japan's experience in importing oil, scrap, iron ore, and coal is representative of the terms that other expanding industrial countries with good port facilities could have obtained during recent years for most material inputs had they chosen to invest heavily in steelmaking.

Japan is chosen because it was obviously the most successful country in exploiting the new economics of steel after 1958. Its output grew steadily from 1950 until the 1958 world recession, but even in the latter year it accounted for only a little more than 4 percent of the non-Communist world output of raw steel. By 1961, however, its steel industry had grown so rapidly that it produced 8 percent of the free world output, and by 1966 its share had increased to 10 percent. (See tables 2-4 and 2-5.)

It was during the late 1950s and early 1960s that one would have expected Japan to begin to have a significant impact on world export markets. Fueled by declining production costs, its exports grew at a rate of 31 percent a year in the 1959–65 period (table 2-5). During this same period U.S. exports were essentially flat at 1.7 million to 2.5 million tons a year, while European exports grew at less than 5 percent annually. Clearly Japan was becoming an important force in the world steel market.

Japanese production costs started to decline in 1958. A combination of falling prices for materials and increasing labor productivity provided much of the impetus for Japan's export surge. This is best seen by examining the trend in Japanese and U.S. labor costs and the costs of basic materials per ton of steel produced (table 3-1). Japanese costs continued to fall through

1. Tariffs have averaged 6 to 7 percent of the value of imports for the past decade.

Table 3-1. *Japanese and U.S. Costs of Finished Steel: Basic Materials and Labor, 1956–76*
Costs in dollars per net ton

	Japan			United States			
Year	*Total basic cost*	*Basic-material cost*[a]	*Labor cost*	*Total basic cost*	*Basic-material cost*[a]	*Labor cost*	*Ratio of Japan/U.S. material cost*
1956	108.72	84.52	24.20	100.55	50.96	49.59	1.66
1957	120.85	96.54	24.31	99.79	45.14	54.65	2.14
1958	89.49	62.17	27.32	110.84	47.26	63.58	1.32
1959	81.68	58.99	22.69	103.40	42.92	60.48	1.37
1960	77.18	56.31	20.87	109.03	43.86	65.17	1.28
1961	83.09	63.19	19.90	111.13	45.49	65.64	1.39
1962	73.99	52.13	21.86	107.72	42.98	64.74	1.21
1963	71.70	50.14	21.56	105.24	42.08	63.16	1.19
1964	68.22	49.20	19.02	104.30	43.52	60.78	1.13
1965	69.29	49.23	20.06	102.50	43.48	59.02	1.13
1966	65.19	46.43	18.76	102.70	42.89	59.81	1.08
1967	63.08	45.00	18.08	106.78	43.38	63.40	1.04
1968	61.49	42.59	18.90	108.32	44.50	63.82	0.96
1969	63.44	44.21	19.23	113.63	45.42	71.21	0.97
1970	70.81	49.74	21.07	124.49	51.18	73.31	0.97
1971	73.74	48.35	25.39	132.43	55.29	77.14	0.87
1972	75.81	46.80	29.01	140.71	59.50	81.21	0.79
1973	91.60	59.56	32.04	146.25	67.04	79.21	0.89
1974	133.63	94.98	38.05	195.55	104.00	91.55	0.91
1975	144.48	99.18	45.30	245.19	124.65	120.54	0.80
1976	146.90	101.87	45.03	267.30	137.08	130.22	0.74

Source: Based on data in U.S. Federal Trade Commission, *The United States Steel Industry and Its International Rivals: Trends and Factors Determining International Competitiveness* (Government Printing Office, 1978), table 3-1, p. 113.

a. Basic materials include only iron ore, scrap, coal, oil, natural gas, and electricity.

Table 3-2. *Ratio of Import Prices (CIF) to Domestic Steel Prices in the United States, 1956–76*

Year	*Hot-rolled sheet*	*Cold-rolled sheet*	*Bars*	*Plate*	*Structurals*
1956	1.42	1.54	1.29	1.66	1.54
1957	1.43	1.25	1.05	1.62	1.36
1958	1.06	1.05	0.73	1.04	1.11
1959	1.15	1.56	0.80	1.05	0.99
1960	1.28	1.54	0.92	1.07	1.12
1961	1.06	1.13	0.82	1.10	1.02
1962	1.00	1.08	0.77	1.13	0.97
1963	0.95	1.00	0.77	1.08	0.92
1964	0.92	0.98	0.78	0.92	0.87
1965	0.91	0.97	0.78	0.94	0.91
1966	0.87	0.92	0.75	0.85	0.87
1967	0.88	0.90	0.75	0.84	0.88
1968	0.85	0.90	0.74	0.77	0.84
1969	0.89	0.88	0.91	0.80	0.86
1970	1.01	0.97	0.88	0.93	0.95
1971	0.98	0.95	0.81	0.85	0.86
1972	1.04	0.97	0.83	0.87	0.86
1973	1.13	1.10	1.00	0.97	1.00
1974	1.25	1.25	1.24	1.24	1.31
1975	1.17	1.10	1.17	1.08	1.10
1976	0.95	0.93	0.86	0.80	0.85

Source: See appendix tables A-3 and A-9.
CIF = Cost including insurance and freight.

1968, a decade during which U.S. costs were essentially constant. Most of the impetus for the Japanese cost decline was fed by savings in the cost of materials, not a decline in labor costs. Between 1957 and 1968 the cost of basic materials fell $54 per ton of steel while unit labor costs fell only $5 a ton. It may be presumed that other exporting countries were enjoying similar declines in materials prices if they imported most of these inputs.

The Model of Prices and Import Penetration

During the late 1950s and 1960s the percentage of world steel output exported to other countries began to rise. The cost of production from modern coastal plants was falling. Competition was increasing as the U.S.-European dominance of the world market eroded. Export prices, and therefore U.S. import prices, began to reflect the efficient producers' costs and changes in demand conditions. (See table 3-2.) In the older producers' markets one would

Figure 3-1. *Relative Prices of U.S. and Japanese Hot-Rolled Sheet, 1956–76*

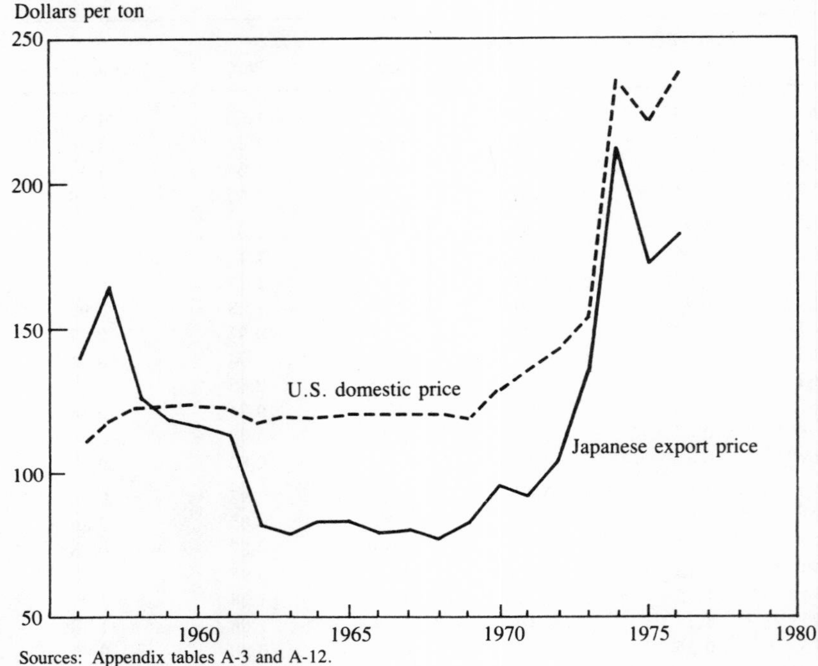

Sources: Appendix tables A-3 and A-12.

expect import competition to begin to have an influence. Either import shares would rise, domestic prices would fall, or both. In the United States an oligopolistic industry could be expected to begin to react slowly, responding to loss of market share by slowing its rate of price increase in a period of mild general inflation. Import shares for most product markets would be expected to rise with a lag. Figures 3-1 and 3-2 provide a rather typical picture of the confluence of these forces for one product, hot-rolled sheet.

The simple theory tested in this chapter, therefore, is that export prices in Japan responded to cost conditions and world demand influences during the 1950s and 1960s. The decline in prices translated into a fall in U.S. import prices after the addition of tariffs and generally declining real freight rates. U.S. domestic prices began to become more sensitive to both import prices and domestic demand as producers fought to stop the erosion of their market shares. Import penetration rose in response to declining relative import prices for steel products, and U.S. capacity ceased to grow as margins were squeezed.

Figure 3-2. *Share of Imported Hot-Rolled Sheet in the U.S. Market, 1956–76*

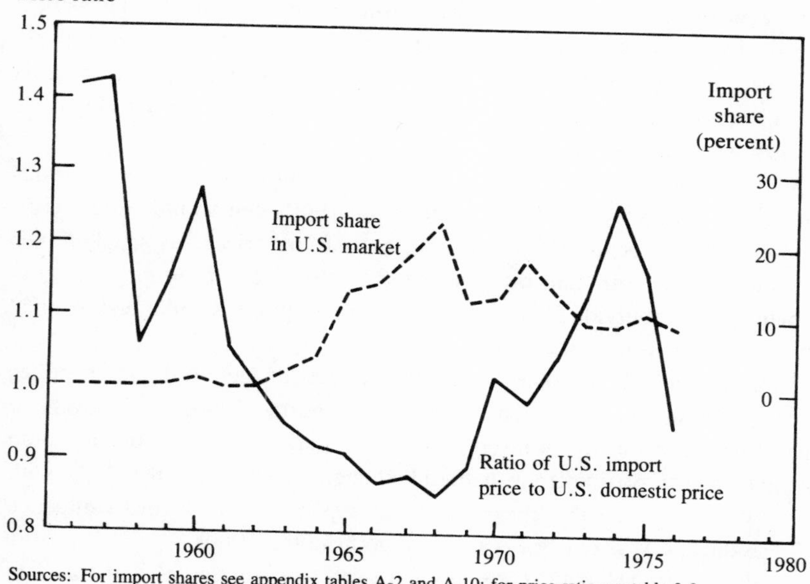

Sources: For import shares see appendix tables A-2 and A-10; for price ratio see table 3-2.

Schematically, these conditions provide a recursive model of U.S. price behavior and import share:

A. $$\begin{bmatrix}\text{World}\\ \text{materials}\\ \text{costs}\end{bmatrix}_t + \begin{bmatrix}\text{Japanese}\\ \text{labor}\\ \text{costs}\end{bmatrix}_t + \begin{bmatrix}\text{World}\\ \text{demand}\\ \text{conditions}\end{bmatrix}_{t-x} \rightarrow \begin{bmatrix}\text{Japanese}\\ \text{export}\\ \text{prices}\end{bmatrix}_t$$

B. $$\begin{bmatrix}\text{Japanese}\\ \text{export}\\ \text{prices}\end{bmatrix}_t + \begin{bmatrix}\text{Freight}\end{bmatrix}_t + \begin{bmatrix}\text{U.S.}\\ \text{duties}\end{bmatrix}_t \rightarrow \begin{bmatrix}\text{U.S.}\\ \text{import}\\ \text{prices}\end{bmatrix}_t$$

C. $$\begin{bmatrix}\text{U.S.}\\ \text{import}\\ \text{prices}\end{bmatrix}_t + \begin{bmatrix}\text{U.S.}\\ \text{costs}\end{bmatrix}_t + \begin{bmatrix}\text{U.S.}\\ \text{demand}\\ \text{conditions}\end{bmatrix}_{t-x} \rightarrow \begin{bmatrix}\text{U.S.}\\ \text{producer}\\ \text{prices}\end{bmatrix}_t$$

D. $$\left[\frac{\text{U.S. import prices}}{\text{U.S. producer prices}}\right]_t \rightarrow \left[\frac{\text{U.S. imports}}{\text{U.S. consumption}}\right]_{t+x}$$

The relationship between Japanese costs, changes in world demand, and Japanese export prices (A) suggests the operation of competitive world market forces. A popular theory among observers in the United States is that the Japanese industry grew by government directive and subsidy and that surges of its exports reflected the desire to subsidize employment during cyclical swings in the Japanese economy.[2] If this latter theory were valid, one would not expect Japanese export prices to track their basic material and labor costs or to vary with changes in world market conditions as much as with Japanese domestic steel consumption.

Since Japanese exports became much more significant in the early 1960s than in previous years, at some point in the 1960s one would expect U.S. import prices to begin to mirror Japanese export prices, as in B. Given declining freight costs and the absence of discrimination by the U.S. government against individual exporters, U.S. import prices should tend to equal Japanese export prices plus freight and duty.

The reaction of U.S. domestic producer prices to changes in import prices for steel (C) should have been increasing during the 1960s. U.S. producer prices historically have followed a cost-plus-markup pattern,[3] but it would not have been surprising if this pattern had begun to change in the 1960s with the growth of imports. Therefore, import prices should have begun to influence U.S. producers' realized prices at some point in the 1960s.

Finally, relationship D represents the confluence of forces that translate changes in the relative prices of imported and domestic steel into lagged changes in market share. As import prices fall, domestic producers respond sluggishly in restraining their own prices. Customers shift after a lag to imported steel, and domestic firms begin to postpone investment plans. This in turn reduces the firms' ability to supply the market at the restrained domestic price. The combination of supply and demand forces may take as little as one year to reach equilibrium (for simple products produced in electric furnaces) or as much as three years or more for the advanced products such as cold-rolled sheet or tin plate. The time lag depends not only on the ability of producers to adjust their product mix but on the time required for fabricators

2. Putnam, Hayes and Bartlett, Inc., *The Economics of International Steel Trade: Policy Implications for the United States* (Newton, Mass.: Putnam, Hayes and Bartlett for the American Iron and Steel Institute, May 1977); and Putnam, Hayes and Bartlett, *The Economic Implications of Foreign Steel Pricing Practices in the U.S. Market* (PB and H for the AISI, August 1978).

3. Richard Rippe, "Wages, Prices and Imports in the American Steel Industry," *Review of Economics and Statistics,* vol. 52 (January 1970), pp. 34–46; and James M. Jondrow, "The Effects of Trade Restrictions on Imports of Steel," in U.S. Department of Labor, Bureau of International Labor Affairs, *The Impact of International Trade and Investment on Employment* (Government Printing Office, 1978), pp. 11–25.

to recognize the existence of lower-cost substitutes of equal quality and to develop delivery and quality-control arrangements so as not to interrupt sophisticated stamping or drawing processes.[4]

The Empirical Model

To test the model of price and market-share determination it is necessary to specify a stochastic form of each equation (A, B, C, and D) that reflects all the important exogenous influences. For instance, there must be some attention paid to the effects of the 1969–74 Voluntary Restraint Agreements (VRAs), the 1974 speculative surge of demand, and the years in which the U.S. industry negotiated labor contracts. It is useful to proceed through a description of each equation.

Japanese Export Prices

The Japanese export price, *JAPPRICE,* is assumed to be driven by labor costs, material costs, and world demand. For material costs, *JMATCOST,* I use the Federal Trade Commission's compilation of basic-material costs for all steel products—coal, oil, iron ore, scrap, and electricity per ton of finished product. For the labor variable, *JLABORCOST,* I use Japanese hourly employment costs, *JPLABOR,* multiplied by e^{-gt}, where g is the annual rate of productivity growth and t is time measured in years beginning with 1955.[5] The value of g is that which minimizes the residual sum of squares in the estimated price equation. The world demand variable is *DEVTREND,* the annual proportional deviation from the long-run exponential growth in free world steel production, 1950–76. Finally, dummy variables are used to capture the effects of the extraordinary speculative demand for steel in 1974 and any shift in the price equation in the early 1960s as Japanese exports began to rise

4. See Helen B. Junz and Rudolf R. Rhomberg, "Price Competitiveness in Export Trade among Industrial Countries," *American Economic Review,* vol. 63 (May 1973), pp. 412–18, for evidence on the speed of adjustment of market shares to changes in prices.

5. The cost data are drawn from U.S. Federal Trade Commission, *The United States Steel Industry and Its International Rivals: Trends and Factors Determining International Competitiveness* (GPO, 1978), with an adjustment for fringe benefits paid to labor. These costs account for roughly two-thirds of total costs and three-fourths of noncapital costs per Peter Marcus, *Price-Cost Model for Japanese Carbon Steel Industry* (New York: Paine Webber Mitchell Hutchins, Inc., September 1978).

sharply. *D*74 is a dummy variable equal to one in 1974 and zero in all other years. *D*62 takes on values of unity in the years 1962–76 and zero in 1956–61.

The form of the export-price equation estimated is simply

$$(3\text{-}1)\quad JAPPRICE_{it} = a_o + a_1\, JMATCOST_t + a_2\, JPLABOR_t\, e^{-g_i t} + a_3\, D62_t + a_4\, D74_t + a_5\, DEVTREND_t + u_{1it},$$

where i indicates the ith product and t the year. The demand variable, *DEVTREND,* is lagged one year ($t - 1$) for some products. In addition a measure of Japanese capacity utilization is added to the equation and substituted for *DEVTREND* to determine whether Japanese *export* prices respond more to world market conditions or to Japanese domestic market conditions. If the Japanese steel exporters sell in a competitive world market, domestic market conditions should be less relevant than world demand. On the other hand if the Japanese exporters are the most aggressive competitors in an imperfectly competitive world steel market, domestic market conditions should affect their competitive urge. To test for this possibility, Japanese capacity utilization, *CUJAP,* and the ratio of Japanese home-market steel consumption to total capacity, *CON/CAP,* are also included in some variants of the equation.

U.S. Import Prices

The equation relating U.S. import prices to Japanese export prices is exceedingly simple. Import prices, *USIMPRICE,* exclusive of transportation charges and tariffs, are directly related to *JAPPRICE,* with the shift variable, *D*62, for the change in market conditions in the early 1960s. This equation is written:

$$(3\text{-}2)\quad USIMPRICE_{it} = b_o + b_1\, JAPPRICE_{it} + b_2\, D62^{*}_t\, JAPPRICE_{it} + u_{2it}\,.$$

It is included to ascertain whether there are sharp departures of U.S. import prices from general export prices (from Japan) during periods of excess capacity or market tightness. Specifically, I shall want to examine the behavior of these prices during the much-discussed "shortage" period of 1973–74.

U.S. Producer Prices

The equation for U.S. producer prices, *USDOMPRICE,* is similar to that for Japanese export prices. It includes a materials cost variable, *USMATCOST,* reflecting basic materials costs for all U.S. finished steel products in each year; a labor cost variable (*USLABORCOST*) that is a product of the total employment cost per hour of labor, *USPLABOR*, and an exponential productivity growth rate, g; and a demand variable, *CUUS*, which is simply the rate of capacity utilization in the U.S. industry. In every case this variable has its impact after a one-year lag. In addition dummy variables for each of the years (1969–72) in which the VRAs might have been binding are included. These are represented by *D*69, *D*70, *D*71, and *D*72. Finally, the price of imported steel, *USIMPRICET*, including importation charges, is included to capture the effect of competition from imports on U.S. producer prices. It is hypothesized that world prices, reflecting the prices of substitutes for U.S. steel, affect U.S. producer prices with very little lag.

The producer-price equation estimated is linear in the variables above.

$$(3\text{-}3)\quad USDOMPRICE_{it} = c_o + c_1\, USMATCOST_t + c_2\, USPLABOR_t\, e^{-g_i t} + c_3\, D69_t + c_4\, D70_t + c_5\, D71_t + c_6\, D72_t + c_7\, USIMPRICET_{it} + c_8\, CUUS_{t-1} + u_{3it}\,.$$

This equation is important for estimating the effects of two rounds of trade protection in the United States: the 1969–74 VRAs and the 1978–79 trigger prices. In addition the equation permits comparison of productivity growth rates and demand sensitivity with the Japanese results from equation 3-1.

Market Shares

The final equation in the simplified empirical model of the industry is the market-share relationship. The import share of each product market in the United States is assumed to be a function of relative prices, time, and labor disruptions. The precise form of the equation is

$$(3\text{-}4)\quad \log\left[\frac{USIMP_{it}}{USSHIP_{it} + USIMP_{it}}\right] = d_o + \sum_{j=1}^{3} d_j \log\left[\frac{USIMPRICET_{it}}{USDOMPRICE_{it}}\right]_{t-j+1} + d_4\, TIME_t + d_5\, DSTR59_t + u_{4t},$$

where *USIMP* represents total imports, *USSHIP* equals shipments by U.S. producers, *TIME* is a linear time trend equal to 1 in 1956, and *DSTR*59 is a dummy variable for 1959, the year of the major postwar steel strike. A time trend is included to capture the shift in U.S. buyers' demand over the 1956–76 period through a learning process.

Equation 3-4 is a "mongrel" equation, capturing both demand and supply

Table 3-3. *Estimates of the Determinants of Japanese Steel Export Prices (Equation 3-1), 1957–76*

												Summary statistics		
Equation	*Product*	*Period*	*Constant*	*JMATCOST*	*JLABOR-COST*	*g*	*D62*	*D74*	*DEV-TREND*$_t$	*DEV-TREND*$_{t-1}$	*CUJAP*$_t$	R^2	ρ[a]	*Durbin-Watson statistic*
1	Hot-rolled sheet	1957–76	23.47	0.8026 (14.52)	250.7 (21.85)	0.11	−22.96 (20.17)	33.93 (9.79)	47.67 (3.60)	. . .	−0.2544 (1.90)	0.996	−0.651	2.269
2	Hot-rolled sheet	1957–76	−0.4224	0.8147 (13.11)	254.9 (19.62)	0.11	−22.28 (17.05)	33.64 (8.81)	28.48 (2.87)	. . .	. . .	0.995	−0.495	1.933
3	Hot-rolled sheet	1957–76	−11.02	0.7929 (10.28)	252.4 (14.35)	0.11	−21.11 (11.87)	37.72 (8.82)	. . .	. . .	0.1380 (1.13)	0.994	−0.215	1.953
4	Cold-rolled sheet	1957–76	0.5479	0.6936 (3.86)	185.4 (8.75)	0.09	−22.19 (4.26)	52.30 (4.63)	−63.59 (1.33)	. . .	0.4942 (1.02)	0.969	−0.342	2.178
5	Cold-rolled sheet	1962–76	−69.91	1.064 (9.75)	301.4 (16.82)	0.11	. . .	35.86 (7.01)	−24.31 (0.83)	. . .	0.6521 (1.98)	0.997	−0.468	1.837
6	Cold-rolled sheet	1962–76	−12.13	1.072 (7.96)	289.2 (13.22)	0.11	. . .	35.86 (6.22)	27.20 (1.34)	. . .	. . .	0.996	−0.271	2.139
7	Cold-rolled sheet	1962–76	−51.85	1.104 (10.61)	293.9 (17.30)	0.11	. . .	34.72 (7.47)	. . .	. . .	0.4362 (2.32)	0.997	−0.351	1.904
8	Bars	1957–76	−61.61	0.5401 (6.28)	192.50 (23.13)	0.09	. . .	96.98 (15.26)	25.49 (1.11)	. . .	0.7284 (3.12)	0.991	−0.736	2.585
9	Bars	1957–76	3.938	0.5309 (4.84)	183.4 (17.57)	0.09	. . .	96.99 (11.77)	79.30 (3.77)	. . .	. . .	0.985	−0.560	2.239
10	Bars	1957–76	−76.14	0.4945 (6.60)	196.6 (26.71)	0.09	. . .	98.93 (16.37)	. . .	. . .	0.9117 (5.60)	0.991	−0.792	2.520
11	Structurals	1958–76	−15.01	0.8208 (6.51)	169.30 (13.39)	0.09	−9.775 (3.52)	71.83 (10.59)	. . .	56.63 (3.22)	0.3196 (1.86)	0.992	−0.369	2.267
12	Structurals	1958–76	17.63	0.7082 (5.58)	170.6 (11.79)	0.09	−9.762 (3.09)	75.95 (10.99)	. . .	56.20 (2.88)	. . .	0.990	−0.312	2.297
13	Structurals	1957–76	−11.36	0.6905 (5.92)	175.10 (13.18)	0.09	−6.806 (2.23)	82.77 (11.11)	. . .	. . .	0.3057 (1.50)	0.987	−0.508	2.015
14	Plate	1958–76	−49.65	1.379 (10.48)	189.80 (11.50)	0.10	. . .	46.93 (5.60)	. . .	49.04 (2.56)	0.2213 (1.09)	0.988	−0.685	2.580
15	Plate	1958–76	−26.79	1.307 (11.34)	189.10 (11.40)	0.10	. . .	50.49 (6.60)	. . .	49.97 (2.61)	. . .	0.988	−0.646	2.563
16	Plate	1958–76	−46.72	1.209 (8.65)	207.10 (11.21)	0.10	. . .	56.12 (6.10)	. . .	. . .	0.2276 (0.92)	0.982	−0.654	2.264

effects. Changes in import prices affect the U.S. industry's willingness to supply domestic markets. As these prices have fallen with time, U.S. prices have been restrained somewhat (equation 3-3), but the U.S. price premium over import prices has widened. This has had two effects: a shift of buyers toward imported steel and a contraction of domestic supply over what it would have been. The total effect of these relative-price changes is thus a combination of demand and domestic supply effects. For this reason the coefficients $d_1 \ldots d_3$ should not be interpreted as "demand" elasticities.

The Empirical Results

The set of equations, 3-1 through 3-4, is estimated for five major product categories: hot-rolled sheet, cold-rolled sheet, bars, structurals, and plate. These constitute approximately 60 percent of U.S. consumption of carbon steel mill products and are the most homogeneous product classes. Other products are either unimportant (rails, tie plates, axles); too heterogeneous (pipe and tubes or wire products); or are produced principally by small electric furnace manufacturers (wire rod). The source of annual data for costs, prices, shipments, imports, and world production may be found in appendix A. All price data, however, reflect *realized* prices, not list prices. Most equations are estimated for the 1956–76 period, using annual data.

Japanese Export Prices

The explanation of export prices for steel from Japan is crucial to any analysis of the U.S. import problems. Do Japanese prices respond to changes in costs and world demand or are they instruments of a Japanese full-employment policy? Equation 3-1 addresses this question. The variable for world demand is the proportionate deviation of each year's world steel production from its long-run trend since 1950. The measure for Japanese demand conditions is simply an estimate of capacity utilization in the Japanese industry, *CUJAP*. If the Japanese cut export prices to maintain employment, one would expect *CUJAP* to assume a positive coefficient in equation 3-1.

Japanese export prices exhibit a strong relationship to material and labor costs throughout the 1956–76 period analyzed. (See table 3-3.) The estimated annual rate of productivity growth, g, is 11 percent for the sheet products, 10 percent for plate, and 9 percent for the bars and structurals. This corresponds very well to the reported average rate of productivity growth for the entire industry of 10.5 percent a year. The coefficients for the labor variable may be interpreted by calculating e^{-gt} for $g = 0.09$ and $g = 0.11$ for each year. For 1976 t is equal to 22; hence the coefficients in table 3-3 must be multiplied by 0.138 and 0.089 to determine the role of labor costs in export prices in 1976. For instance, the coefficient for hot-rolled sheet translates into twenty-two times the Federal Trade Commission's reported measure of 1976 hourly compensation. Since this latter number may be as much as 20 percent below the hourly labor costs for the integrated Japanese firms, and hot-rolled sheet requires about six man-hours a ton to produce and ship, the final interpretation is that Japanese export prices reflect basic-material costs plus three times the cost of the labor used in the production of this product. (See appendix tables A-12 and A-17.) The latter charge obviously must cover labor, overhead, and nonbasic materials costs.

The differences in the materials cost coefficients reflect the variance in energy use and yields across products. Bar products are likely to require less energy and result in higher yields than sheet products; hence the coefficients of *JMATCOST,* the average materials cost across *all* products, are lower for the bar products. The coefficients of *JMATCOST* for cold-rolled sheet are unreasonably low for 1957–76 but rise sharply for the 1962–76 regressions. The other coefficients change rather drastically as well, reflecting a structural shift in Japanese export pricing policy for this product in the early 1960s.

In general the proxy for world-market demand, *DEVTREND,* performs fairly well in the price equation, but Japanese capacity utilization, *CUJAP,* also enters significantly in some product equations. For bars and cold-rolled sheet, *CUJAP* appears to exert the more systematic influence, but for hot-rolled sheet, structurals, and plate, the world demand variable clearly dominates *CUJAP* when both are in the same equation. This is consistent with the view that Japanese export prices are driven by world market forces and to a somewhat lesser extent by Japanese demand conditions. The coefficients of *CUJAP* are very small, however, suggesting that the Japanese do not slash export prices dramatically when capacity utilization falls. The largest coefficient, 0.91, predicts a decline of only $9 in bar prices for a decline in 10 percentage points in capacity utilization.

In general, small values of *DEVTREND* have very little influence on export prices, but a nonlinear formulation is successful only for cold-rolled sheet. While this formulation outperforms *DEVTREND* for cold-rolled sheet, it does not perform as well in the price equation for the other products and is therefore not reported. When values of *DEVTREND* less than 0.05 are excluded, the fit of the equations improves somewhat for most of the products, confirming the nonlinear effect of demand changes on price. The differences are so slight, however, that this latter specification is not reported in table 3-3.

It is possible that the Japanese cut export prices to increase employment in periods when capacity is not being fully utilized to satisfy home-market requirements. If this is true, the addition of a variable measuring the ratio of Japanese domestic steel consumption to domestic capacity (*CON/CAP*) should perform better than the capacity utilization ratio. In fact, this does not occur. (See appendix table B-2.) When this variable is subtituted for *CUJAP*, the explanatory power of the estimated equations declines for the sheet products, remains virtually the same for bars, and improves slightly for structurals. World market conditions continue to exert a more powerful influence on structural prices than do home-market consumption levels.

In summary, it is clear that current costs drive Japanese export price realizations and have done so since 1956 or 1957, when Japan was exporting very small amounts of steel. The effect of demand forces is much milder, however, as table 3-3 shows. The impact of deviations from the long-term trend in world production affects export prices, but modestly. The estimated coefficient of *DEVTREND* implies that an increase in world production levels that would reduce the proportional deviation from trend from -0.15 to -0.05 would increase prices by $2.50 to $8 a net ton. Given an average price of $120 for the five products over 1956–76, this translates into a change of 2.0 to 6.6 percent for a major shift in world demand. In addition to this effect, however, there is undoubtedly an indirect effect of world demand on price through its impact on materials prices. Therefore, while Japanese steel demand influences export prices, particularly for bar products, these influences are less systematic and generally less powerful than world demand forces.

The dummy variable, *D*62, takes a value of unity for the years 1962–76 and zero before 1962. It is included to capture the effect of the Japanese emergence as competitors in the world market. For hot-rolled sheet and structurals Japanese export prices are lower after 1961, but this effect is less systematic for other products.

The dummy variable *D*74 is equal to unity in the year of near-panic buying of steel, 1974, and zero otherwise. There is no other simple way to capture the effect of this one-year meteoric rise in steel prices. In virtually every case the coefficient of *D*74 is large and significant, ranging from $34 to $97 a ton.

Since there is a decided shift in structure in many of the export price equations in 1962, alternative estimates of equation 3-1 for 1962–76 are presented in appendix B. These estimates are used in subsequent analyses of U.S. import policy.

Table 3-4. *Residuals from Japanese Export Price Equation 3-1*[a]
Dollars per ton

	Product				
Year	*Hot-rolled sheet (Equation 1)*	*Cold-rolled sheet (Equation 4)*	*Bars (Equation 8)*	*Structurals (Equation 11)*	*Plate (Equation 14)*
1957	−0.60	−6.79	3.00		
1958	−0.08	−4.22	0.62	12.86	−4.02
1959	0.75	−0.84	0.07	0.68	−5.46
1960	1.73	19.46	6.91	3.61	8.65
1961	−1.20	−9.34	−1.21	−5.68	−0.56
1962	−2.51	−2.31	−4.27	1.56	−6.04
1963	−0.32	5.51	2.22	2.94	−2.37
1964	0.45	7.54	−6.28	−0.90	2.44
1965	0.06	5.69	−2.51	−4.11	5.77
1966	0.72	0	1.32	−2.73	1.39
1967	3.85	−4.50	4.03	5.26	1.38
1968	1.49	−5.04	4.76	4.17	−0.29
1969	−4.05	−4.92	−7.12	−1.49	−4.29
1970	−0.58	4.23	−0.04	5.33	−0.33
1971	−0.06	−7.49	2.44	−4.51	−1.98
1972	−1.26	−6.46	−5.42	−5.60	4.04
1973	0.09	3.93	6.28	2.52	0.08
1974	2.69	0.68	−1.77	−0.96	−7.00
1975	−4.14	−1.98	2.40	2.59	10.22
1976	2.58	4.52	−3.23	−1.37	−4.39

Source: Author's estimates.
a. Equation numbers in headings are from table 3-3.

The strength of the results of equation 3-1 suggests that the prices of each product respond to competitive forces. Only cold-rolled sheet prices failed to respond to world demand changes over the 1956–76 period. There are no products for which prices broke sharply from the trend in production costs over any prolonged period. There is no period of persistently large negative residuals from the equations' predictions (see table 3-4) that might signal a concerted effort by the Japanese to use price cutting as a technique for expanding market share. The negative residuals for sheet products in the 1960s are generally quite small, averaging less than $4 per net ton shipped, surely not enough to lure large quantities of new business. In short, it is difficult to reject the hypothesis that it was sharply declining relative costs in Japan (and other countries) that fed the export surge in the 1960s and drove world steel prices down in current dollars for most of the 1960s.

A graphic illustration of the difference between the behavior of domestic producer prices and of Japanese export prices is found in figure 3-3. The

Figure 3-3. *Price-Cost Margins for Japanese Exports and U.S. Producers' Domestic Sales, 1956–76*[a]

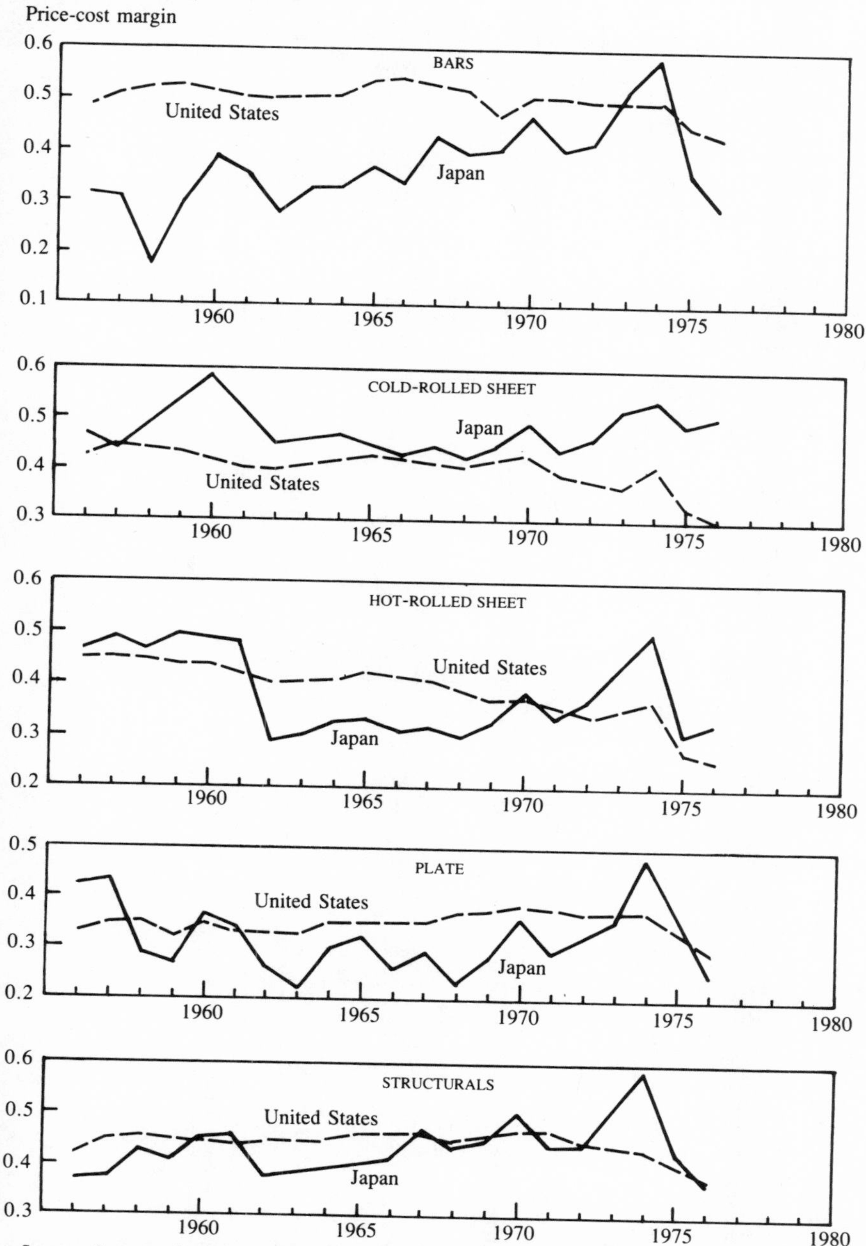

Sources: See appendix tables A-15 and A-16 for costs, A-3 and A-12 for prices.

a. Price-cost margin is average realized price minus labor and basic-material costs divided by average realized price.

Table 3-5. *Determinants of U.S. Import Prices (Equation 3-2), 1957–76*

Product	Period	Constant	JAPPRICE	D62*JAPPRICE	Summary statistics		
					R^2	ρ^a	Durbin-Watson statistic
Hot-rolled sheet	1957–76	−3.096	0.9222	0.2042	0.949	0.046	2.009
			(13.64)	(4.61)			
Cold-rolled sheet	1956–76	−10.90	1.068	. . .	0.901	0	1.791
			(13.57)				
Bars	1957–76	−16.05	1.209	. . .	0.903	0.138	1.931
			(11.40)				
Structurals	1956–76	−14.07	1.133	. . .	0.941	0	2.135
			(17.85)				
Plate	1957–76	3.950	0.9397	0.1431	0.973	0.450	1.712
			(12.41)	(2.54)			

Source: Author's estimates. Figures in parentheses are *t*-statistics.
a. Hildreth-Liu correction for serial correlation.

price-cost margins for Japanese exports and U.S. domestic sales are computed for each of the five major products, utilizing a vector of input coefficients based on 1976 efficient technology and trended labor productivity growth. It is important to stress that this measure of costs includes only basic materials and labor and that the input coefficients do not vary with the cycle. (See appendix tables A-15 and A-16 for the detailed estimates.) The price-cost margin should be thought of as a margin over ideal or ''optimal'' basic-material and labor costs, not actual costs.

In every case the Japanese export price-cost margin varies more with cyclical fluctuations than does the U.S. series. This is because *prices* vary with world cycles in the export market. Except for bars, the Japanese price-cost margins in the 1970s are as great as those in the United States. Put another way, Japanese markup over efficient costs is as great as the U.S. markup even though Japanese firms are likely to be more efficient because of newer plants in the later years. In the early 1960s the Japanese margins were smaller than their U.S. counterparts, but these differences eroded steadily in the late 1960s and early 1970s. Despite this phenomenon, it was only after 1969 that charges of ''unfair'' trade practices began to be levied against the Japanese.

Obviously the series portrayed in figure 3-3 cannot be used to evaluate charges of dumping under the 1974 Trade Act cost-of-production standard. Economic evidence is largely irrevelant in these dumping allegations (see chapter 5). But the penetration of Japanese steel into the U.S. market in the 1970s would seem to have been quite profitable, given the upward trend in price-cost margins until 1975. Of course, after 1975 the world market was so weak that it would require a cartel to make the industry look profitable, a deduction not lost on those pleading for special steel-sector negotiations within the Organization for Economic Co-operation and Development.

U.S. Import Prices

Table 3-5 contains the results of estimating the determinants of U.S. import prices. Not surprisingly, U.S. import prices (excluding freight, handling, and duty) respond directly and immediately to Japanese export prices. They are simply price observations in the same market. For two of the products, hot-rolled sheet and plate, there is a shift in the price-price relationship in 1962, but not for the other three.

The direct link between world export prices (as measured by Japanese export price realizations) and U.S. import prices was weakened substantially in 1973. Beginning in the spring of 1973, export prices of steel rose sharply, more than doubling by late 1974. But during this period the United States was in the midst of a bout with price controls. Phased decontrol began in the spring of 1974, more than one year after the beginning of the "steel shortage" of 1973–74.

The effects of price controls on U.S. steel import prices was dramatic. While imported materials were exempt from price controls, buyers must have had difficulty passing on the sharply rising steel costs in their fabricated products. Put another way, the derived demand price for imported steel was restrained by price controls on fabricated products. As world export prices soared, U.S. import prices lagged far behind, and steel imports into the United States declined from 17.7 million tons in 1972 to 15.2 million tons in 1973 (table 2-4). The reason for this decline is easily seen in the 1973 residuals from the import-price equations (equation 3-2) reproduced in table 3-6. These residuals are consistently negative and the largest in absolute value of any year in the 1956–76 sample period. U.S. import prices were from 1 percent (plate) to 15 percent (bars) below predicted values. As a result, imports fell, but not by much. Despite the low prices in the United States, importers were able to obtain 85 percent of the 1972 import supply.

This pernicious effect of price controls has been lost on other students of the industry who argue that the 1973–74 "shortage" experience demonstrates the need for greater domestic self-sufficiency in steel. They argue that in times of shortage, exporters abandon the U.S. market.[6] But the results in table 3-6 allow one to conclude that this contraction of imports was probably mostly due to a government-induced failure of U.S. importers to bid for steel on the rising world market. In fact, the low U.S. prices led American firms to increase their exports by more than foreign exporters contracted their shipments to the United States. Imports declined by 2.5 million tons between 1972 and 1973 but then rose by 0.8 million tons in 1974. Exports rose by 1.2 million tons in 1973 and by another 1.8 million tons in 1974, a year of extraordinarily high world steel prices.

U.S. Producer Prices

The U.S. producers' prices, as measured by unit values reported annually to the Department of Commerce (*USDOMPRICE*), follow basic materials and labor costs in the same year much as do the Japanese export prices (table 3-

6. Putnam, Hayes and Bartlett, *The Economic Implications of Foreign Steel Pricing Practices in the U.S. Market*.

Table 3-6. *Residuals from Equation 3-2 during U.S. Price Controls, 1972–74*[a]
Percent

Product	1972	1973	1974
Hot-rolled sheet	6.9	−8.5	−1.2
Cold-rolled sheet	5.5	−9.0	3.9
Bars	5.5	−15.4	−6.3
Structurals	6.8	−12.7	−6.8
Plate	2.8	−1.0	−2.4

a. Residual equals actual price minus predicted price divided by predicted price times 100.

7). The materials-cost coefficients exhibit the expected pattern of increasing as the sophistication of the product increases, except for plate, which has a rather low coefficient given the yield losses in producing this product. The labor coefficients exhibit the expected progression with increasing sophistication of product, and in every case except structurals and bars the best fit occurs with a rate of productivity growth of only 1 percent a year. The structurals and bars equations' residual sum of squares is minimized with a hypothesized 2 percent rate of productivity growth. Recall that the Japanese magnitudes were 9 to 11 percent over the same period.

Except for structurals, many of which are produced by small firms, steel product prices respond to changes in U.S. capacity utilization, generally with a one-year lag. A shift in capacity utilization from 90 to 95 percent, for example, would raise prices of sheet products by about 60 cents a net ton in the following year, or less than 0.5 percent on the average over the period. Bars would rise by somewhat more, but less than 1 percent, assuming a similar shift in utilization. Thus U.S. prices are sensitive to changes in demand but less so than world prices.

The effect of import prices on U.S. producers' realized prices is felt in the same year. An increase in world steel prices of $1 raises U.S. domestic prices by 6 cents to 11 cents, depending on the product. Since the United States exports very little steel—about 2 to 3 million tons annually—it could hardly be argued that the causation runs from U.S. prices to world prices. It must be that world market conditions affect U.S. prices with very little lag. The *t*-statistics for both the import-price and demand variables are generally greater for the 1962–76 period than for the entire 1957–76 or 1958–76 periods, particularly for the sheet products (see table B-3). This is consistent with the hypothesis of increasing competition in the early 1960s due to the emergence of the Japanese as a major competitive force.

Table 3-7. *Determinants of U.S. Producer Prices (Equation 3-3), 1958–76*

Product	Period	Constant	USMAT-COST	USLABOR-COST	g	D62	D69	D70	D71	D72	USIM-PRICET	$CUUS_{t-1}$	Summary statistics R^2	Summary statistics ρ^a	Summary statistics Durbin-Watson statistic
Hot-rolled sheet	1958–76	47.57	0.8953	3.219	0.01	. . .	−4.517	−5.007	−0.09381	−3.395	0.1149	0.08616	0.999	−0.887	2.31
			(13.45)	(3.12)			(3.00)	(2.93)	(.06)	(1.47)	(9.30)	(2.76)			
Hot-rolled sheet	1958–76	48.27	0.9807	1.832	0.01	. . .	−3.855	−4.199	. . .	. . .	0.1108	0.1104	0.999	−0.894	2.45
			(26.81)	(3.60)			(3.04)	(3.32)			(9.47)	(4.14)			
Cold-rolled sheet	1958–76	39.72	1.041	5.895	0.01	5.812	−5.387	−0.09126	3.251	2.561	0.1060	0.1406	0.999	−0.917	2.188
			(14.60)	(4.53)		(5.28)	(3.58)	(−0.05)	(2.08)	(1.04)	(9.37)	(4.61)			
Cold-rolled sheet	1958–76	39.33	0.9778	7.110	0.01	5.294	−4.453	. . .	3.014	. . .	0.1035	0.1248	0.999	−0.899	2.097
			(24.57)	(10.48)		(5.44)	(3.57)		(2.31)		(9.43)	(4.55)			
Bars	1958–76	12.66	0.8008	17.14	0.02	. . .	−19.00	−5.75	0.4307	2.396	0.09836	0.3094	0.997	−0.730	2.215
			(4.96)	(4.59)			(5.70)	(1.58)	(0.12)	(0.47)	(4.11)	(4.89)			
Bars	1958–76	10.16	0.7334	18.88	0.02	. . .	−19.63	−6.16	. . .	. . .	0.09663	0.2965	0.998	−0.688	2.156
			(8.20)	(10.56)			(7.28)	(2.24)			(4.35)	(5.26)			
Structurals	1957–76	−1.772	0.5971	23.62	0.02	. . .	−0.8233	−2.249	6.024	1.966	0.05946	0.0974	0.998	0	1.895
			(5.01)	(10.29)			(0.33)	(0.92)	(2.45)	(0.71)	(2.51)	(1.59)			
Structurals	1957–76	−3.989	0.5514	24.57	0.02	. . .	. . .	. . .	5.496	. . .	0.06182	0.1024	0.998	0	2.095
			(5.85)	(14.47)					(2.50)		(2.79)	(1.89)			
Plate	1958–76	7.282	0.5687	18.95	0.01	−1.963	−3.447	−0.1144	5.333	2.666	0.07492	0.1482	0.999	−0.643	2.523
			(5.78)	(10.27)		(1.55)	(1.50)	(0.05)	(2.33)	(0.78)	(4.83)	(3.17)			
Plate	1958–76	7.878	0.5825	19.32	0.01	. . .	. . .	. . .	4.475	. . .	0.08147	0.1209	0.999	−0.277	2.094
			(7.70)	(18.02)					(1.96)		(4.47)	(2.47)			

Source: Author's estimates. Figures in parentheses are *t*-statistics.
a. Hildreth-Liu correction for serial correlation.

Finally, the effects of the VRAs of 1969–72 are estimated by including dummy variables for each year. The most pronounced effects occur in 1971 for plate, structurals, and cold-rolled sheet. Hot-rolled sheet and bars evidence negative coefficients for *D*69 and *D*70 and nonsignificant coefficients for *D*71 and *D*72. The results generally confirm the view that the VRAs had an increasing impact on domestic prices as they were implemented from 1969 through 1971, but their impact declined with the imposition of price controls in late 1971.

To summarize, the producer-price equations, based on annual observations of realized prices, demonstrate that cost and demand conditions affect U.S. prices, the former immediately and the latter generally after a one-year lag. In fact, the response to cost changes appears to be spread out over several years, but the inclusion of lagged cost variables has little impact on the standard error of the estimate. Given problems of collinearity, only the current-year costs are included in the equations. It is clear that import competition began to have an effect on domestic prices roughly in 1962, and it was in this year that U.S. capacity utilization began to feed back on prices. Thus these results are not inconsistent with the Stigler-Kindahl results, which showed little price cutting during the 1958 and 1960–61 recessions.[7] Finally, the impact of the VRAs appears to have been very uneven, a subject to which I return in chapter 5.

Market Shares

Table 3-8 shows the effect of lagged relative import prices on the import penetration of each product. A dummy variable is included to capture the effect of the 1959 strike and a time trend (*TIME*) captures the shift in share over the 1957–76 period, perhaps reflecting better information about alternative sources of supply. The coefficients of the relative-price terms vary across the five products, generally increasing with the value of the products. The simple products, bars and structurals, exhibit the shortest lags and the lowest relative price coefficients undoubtedly because small, flexible minimills are important suppliers and because considerable geographical specialization takes place in these markets. Cold-rolled sheet, on the other hand, exhibits longer lags of up to two years and large relative price coefficients. Buyers adjust, but only after considerable lags. Similarly specialized cold-

7. George J. Stigler and James K. Kindahl, *The Behavior of Industrial Prices* (New York: National Bureau of Economic Research, 1970), pp. 71–75.

Table 3-8. *Determinants of Import Market Share (Equation 3-4), 1957–76*[a]

Product	Period	Constant	Log (relative import price)[b]			TIME	DSTR59	Summary statistics		
			Current	Lagged 1 year	Lagged 2 years			R^2	ρ^c	Durbin-Watson statistic
Hot-rolled sheet	1958–76	−5.270	. . .	−7.069 (6.67)	. . .	0.1930 (6.96)	. . .	0.881	0.061	1.938
Cold-rolled sheet	1959–76	−5.206	−1.463 (1.92)	−1.820 (1.80)	−4.147 (4.96)	0.1888 (10.15)	. . .	0.954	−0.271	2.067
Bars	1958–76	−4.496	. . .	−1.743 (3.38)	. . .	0.1043 (5.01)	. . .	0.846	0.514	1.750
Structurals	1957–76	−2.571	−1.365 (4.27)	. . .	. . .	0.06114 (7.20)	0.6215 (3.35)	0.868	0.139	1.980
Plate	1957–76	−4.235	−1.431 (1.92)	−2.152 (3.18)	. . .	0.1373 (6.92)	0.9991 (2.28)	0.906	0	2.245

Source: Author's estimates. Figures in parentheses are *t*-statistics.
a. Dependent variable: LOG [*USIMP*/(*USIMP* + *USSHIP*)].
b. Relative import price = *USPRICET* ÷ *USDOMPRICE*.
c. Hildreth-Liu correction for serial correlation.

reducing mills cannot be converted to other products (other than to producing cold-rolled sheet for tin plate); hence shifts in domestic supply occur only through failure to maintain cold-reducing mills or in delaying modernization outlays on these mills.

In general the other variables take the expected signs. The 1959 strike increased the import share for two of the products. Moreover, import shares appear to be rising with time regardless of the movement of other variables in the equation, perhaps a reflection of reduced transportation costs and improved communication of price-quality options to buyers. Finally, the import share is not sensitive to changes in the level of demand as measured by industrial production.

The combination of the rather modest response of U.S. producers' prices to changes in import prices and the considerable elasticity of market share with respect to the relative price of foreign steel provides a rather interesting clue to the nature of the domestic and import markets. It would seem that domestic producers do not feel that they have to respond fully to imported price reductions to defend their primary domestic markets. While they have become more sensitive to world market conditions since 1962, they adjust prices by only approximately one-tenth of the movement in imported prices. Despite this fact, importers' market shares seem quite sensitive to price; the elasticity with respect to relative price varies from -1.4 to -7.4 in the long run (table 3-8). This suggests that the portion of the U.S. market that importers penetrate is price sensitive. While these elasticities are not simply demand elasticities, it appears that the U.S. steel market consists of two types of buyers: those desiring long-term arrangements with secure domestic suppliers and those more interested in minimizing their short-term cost of steel. The former are much less sensitive to relative prices than the latter. In fact, estimates of import demand equations by themselves result in own-price elasticities of -2.5 to -10.4 for the sheet products, while domestic demand elasticities for the same products range from 0.5 to 1.5 (see appendix tables B-4 and B-5). These are consistent with the results of Jondrow.[8] On the other hand cross-piece elasticities are substantially lower for domestic steel demand. This latter result is new, although there have been few attempts to estimate these cross-elasticities. In fact, Takacs's study assumes perfect substitutability,[9] an assumption clearly inconsistent with the results of this chapter.

8. Jondrow, "The Effects of Trade Restrictions on Imports of Steel."

9. Wendy E. Takacs, "Quantitative Restrictions on International Trade" (Ph.D. dissertation, Johns Hopkins University, 1975).

Summary: The Relevance of the Results

A number of general conclusions emerge from this chapter. First, it is clear that foreign, and particularly Japanese, costs began to decline relative to U.S. costs beginning in 1958, the year before the U.S. industry took its longest postwar strike. Following this break in relative costs, the price of imports into the United States began to decline relative to U.S. prices. The decline in relative prices was steep for almost a decade, and by 1963 or 1964 virtually all basic carbon steel import prices had fallen below U.S. domestic prices even after including all freight, handling, and import duties.

Second, the Japanese behavior in export markets can best be characterized as extremely competitive. Export prices were not manipulated simply to control domestic employment. Rather, they reflected surely and quickly the change in production costs and changes in demand conditions. Prices responded to both within a year. In the 1960s Japanese export prices of hot- and cold-rolled sheet were frequently below the values projected by the competitive model, but with only a few exceptions these deviations were very small. Similarly, in 1974—the year of booming demand—prices were substantially above projections.

These departures from projected export prices do not seem to follow a predictable pattern except for the 1974 experience. Rather, they represent the temporary departures from equilibrium pricing by sellers in a rapidly changing market. But these departures should not detract from the central conclusion that most of the variance in export prices can be explained by movements in costs and demand. By the early 1960s the market had become very competitive.

Third, while lags are short in translating factor cost changes into relative prices, they are decidedly longer in influencing market shares of domestic and imported steel in the United States. For the simpler products, relative prices affect import shares after a lag of a year or less. But for sheet products the lags run as long as two years. Therefore, the disastrous effects of the 1975 recession were recorded quickly in prices and in the market shares for the simpler products. The market shares of cold- and hot-rolled sheet might not have begun to respond until 1976 or even 1977. For this reason it is not surprising that it was early in 1977 that steel producers began to cry for relief from imports even though the relative price of imports had tumbled by between 10 and 20 percent in 1975 and by another 20 percent in 1976 (table 3-2). Most of the effect of these changes was registered in 1977. Similarly, one might expect the effects of the trigger price enforcement of U.S. dumping laws to take effect in 1979 and 1980. Thus policy initiatives designed for short-term effect must eschew control of prices.

In summary, the results of this chapter provide substantial evidence that competitive market forces have eroded the U.S. producers' share of the price-sensitive portion of the U.S. steel market. This is not a recent phenomenon, nor is it very difficult to explain. It is rooted in a world market in which U.S. producers' relative costs rose consistently for the decade of the 1960s. In part this was due to high and rising labor costs, but a large share of the change in relative prices derived from increasing world competition in the production and shipping of basic raw materials. There was nothing U.S. producers could have done to reverse this trend. They were simply the victims of market and technological forces.

IV

The Economics of Expansion for Basic Carbon Steel

There can be little doubt that part of the U.S. steel industry's problem in competing with the rising tide of imports is its embedded technology. Most U.S. facilities were built before World War II. While they have been modernized to a considerable degree, they are simply less efficient than the most modern new plants. It is difficult to fit new technology—top-blown basic oxygen furnaces (BOFs), large blast furnaces, continuous casters, and energy recovery systems—into old plant layouts without completely suspending production and redesigning the plant site. Most visitors to U.S. plants are surprised at the apparent inefficiency of the materials flow, but they fail to recognize the costs of improving that flow once a plant is built and operating.

In a growing market prices must eventually reflect the cost of production from new capacity. These prices determine the continued viability of old plants since these older facilities will only be kept in production if the marginal cost of producing steel in them is no greater than the average cost of production from new capacity. Thus it is necessary to know the cost of producing steel from new facilities in various countries before one can assess the long-run prospects for the U.S. industry.

Since the enactment of the Trade Act of 1974, which includes a new cost-of-production test for antidumping cases, comparisons of production costs for mills in various countries have become popular.[1] These comparisons are generally based on averages of accounting costs across plants of different

1. See, for example, Peter F. Marcus and Karlis M. Kirsis, *World Steel Dynamics: The Steel Strategist* (Paine Webber Mitchell Hutchins, June 1980); Putnam, Hayes and Bartlett, Inc., *The Economic Implications of Foreign Steel Pricing Practices in the U.S. Market* (Newton, Mass.: Putnam, Hayes and Bartlett for the American Iron and Steel Institute, August 1978); U.S. Council on Wage and Price Stability, *Report to the President on Prices and Costs in the United States Steel Industry* (COWPS, October 1977); and Hans Mueller and Kiyoshi Kawahito, *Steel Industry Economics: A Comparative Analysis of Structure, Conduct, and Performance* (New York: Japanese Steel Information Center, 1978).

vintages and technological efficiency. Unfortunately such comparisons are of little use in predicting the course of prices and the location of production in future years.

In the United States, for example, the net book value of fixed steelmaking assets in 1978 was $100 per net raw ton of capacity.[2] The market value of these assets has hovered around 50 percent of their book value in the past several years, or approximately $50 per net raw ton,[3] but the replacement cost is at least fifteen times this market value. Therefore, it is quite clear that investors do not expect steel producers to be able to recover the full capital charges on the book value of their assets and that they value their assets far below the cost of replacing them.

Unless the variable costs of production from the new facilities are drastically less than the costs of producing from current assets, these new facilities will not be built in the United States. This will mean that U.S. steel prices will eventually be determined by the cost of production in other countries plus the cost of importation. Production from the older plants can and will continue, and some of these plants may even expand, but most of the growth in U.S. steel consumption will be satisfied by foreign suppliers. On the other hand if large cost savings are available from new mills, U.S. prices will eventually reflect production costs from new U.S. mills.

In either case the production costs of existing mills are largely irrelevant to the determination of the long-run equilibrium price. Instead, these costs are important only to investors in determining the value of existing mills and their survival. A comparison of production costs across existing mills is useful, therefore, only for determining which plants will survive and which will close as the world market approaches equilibrium. To determine the long-run equilibrium price requires an analysis of the cost of building new capacity in various parts of the world.

2. *Annual Statistical Report: American Iron and Steel Institute, 1978* (Washington, D.C.: AISI, 1979), pp. 16, 55. Firms reporting to AISI indicated total net fixed assets of $20,272 million. Industry capacity was 157.9 million tons, and reporting firms owned 88.8 percent of this capacity. Steel revenues were 75.6 percent of total company revenues, given shipments of 86.4 million net tons at an average price of $406 a net ton (U.S. Bureau of the Census, *Current Industrial Reports: Steel Mill Products*, Series MA-33B [Government Printing Office, 1978]). Therefore, net fixed assets in the steel industry may be estimated to be approximately 75.6 percent of reported assets, or $15,326 million. This translates into a book value of $109 a net ton, of which approximately 10 percent represents the investment in mining, processing, and shipping raw materials.

3. Standard and Poor's, *Analysts Handbook, 1979* (New York: Standard and Poor's, 1979), p. 73.

The Cost of Building Capacity in the United States

New capacity to produce steel can be obtained in three different ways: (1) by modernizing mills, replacing older facilities with new, larger-scale installations; (2) by adding new facilities to existing plants; and (3) by building new plants.

The modernization process often includes the rounding-out of capacity because past alterations in a plant's design have left the plant out of balance. For instance, replacement of blast furnaces and steelmaking furnaces may provide greater raw steel capacity than existing rolling mills can handle. Therefore, replacement or modernization of a hot-strip mill will round out capacity. The addition of new equipment to existing plants that expands capacity throughout is often referred to as brownfield expansion. Finally, the construction of an entirely new plant, as noted earlier, is described as greenfield expansion.

Since 1960 the U.S. industry has spent most of its capital resources on modernizing plants. Only one new integrated mill (Burns Harbor) has been built in two decades. During the 1960s the Japanese and European industries expanded their capacities substantially through the construction of new plants, and in the 1970s the less developed countries began to construct substantial new integrated capacity. Therefore, there is at least some recent experience on which to base estimates of new-plant (greenfield) costs for most producing areas other than the United States. For the United States, however, the data must be drawn from engineering estimates.

Burns Harbor

In the early 1960s Bethlehem Steel Corporation began the construction of finishing facilities at Burns Harbor, Indiana, on Lake Michigan. By 1966 it had completed an 80-inch hot strip mill, a 160-inch sheared-plate mill, a cold-rolling mill, and a tinning mill.[4] These mills used slabs imported from other mills until the late 1960s, when Bethlehem added basic steelmaking capacity at Burns Harbor. By 1970 it had completed a fully integrated steelworks able to produce 2.5 million to 3.0 million tons of raw steel and to finish a full array of flat-rolled products. The full cost in current dollars (1962 to 1970

4. Details can be found in the annual reports of the Bethlehem Steel Corporation for the years 1966 through 1970.

dollars) was nearly $1 billion, including all site costs.[5] More than half of this expenditure was invested in the later basic steel facilities. In 1969 dollars it would appear that the cost of Burns Harbor was approximately $350 per net raw ton of capacity.

AISI-EPA Study

In 1975 the American Iron and Steel Institute commissioned a study of the cost of environmental controls and the needs for capital expansion.[6] This study was prepared by Arthur D. Little, Inc. (ADL), from industry data. A detailed cost estimate for building 1 million tons of net finished steel capacity, assuming the full mix of carbon products produced by the industry, was derived from this report in a study by Temple, Barker, and Sloane, Inc. (TBS), for the Environmental Protection Agency (EPA).[7] The raw steelmaking capacity required for this 1-million-ton finished mix was estimated to be 1.562 million tons. The cost of all facilities, excluding site preparation, shops, railroads, and engineering and start-up costs, was $408 per net raw ton, an estimate with a potential upward bias given the large percentage of conventional ingot pouring and yield loss assumed. The details appear in table 4-1.

Recent Industry Estimates

A 1978 industry estimate of the cost of a 4-million-net-raw-ton greenfield plant, producing only flat-rolled products, was $3.6 billion, but 27 percent of the cost represents overhead costs that may be spread out over subsequent additions to capacity.[8] As in the ADL-TBS estimate in table 4-1, approximately one-half of the costs are for basic steelmaking, through slabs; the remainder is in finishing costs.

Only United States Steel has mentioned the possibility of building a new facility in recent years. In 1977 Edgar Speer, chairman of U.S. Steel, announced that the company was beginning to make plans to construct a plant

5. Bethlehem Steel Corporation, Burns Harbor facility, press release, January 4, 1979 (year-end statement of the general manager).

6. Arthur D. Little, Inc., *Steel and the Environment: A Cost Impact Analysis,* A report to the American Iron and Steel Institute (Cambridge, Mass.: Arthur D. Little, May 1975).

7. Temple, Barker, and Sloane, Inc., *Analysis of Economic Effects of Environmental Regulations on the Integrated Iron and Steel Industry* (U.S. Environmental Protection Agency, July 1977), 2 vols.

8. Confidential information supplied by a major steel company.

Table 4-1. *Cost of Building a New Integrated Carbon Steel Mill in the United States*
Dollars per net raw ton

Stage of production	Estimated cost		
	ADL-TBS[a] *(1975 dollars)*	*U.S. Steel*[b] *(1976 dollars)*	*Industry sources*[c] *(1978 dollars)*
Material preparation (through the blast furnace)	103	...	248
Steelmaking	32	...	35
Semifinishing	66	...	62
Finishing	206	...	313
Total (excluding site)	408	...	658
Site preparation	...	...	242
Grand total	...	600	900

a. Arthur D. Little, Inc., *Steel and the Environment: A Cost Impact Analysis,* A report to the American Iron and Steel Institute (Cambridge, Mass.: Arthur D. Little, May 1975); and Temple, Barker, and Sloane, Inc., *Analysis of Economic Effects of Environmental Regulations on the Integrated Iron and Steel Industry* (U.S. Environmental Protection Agency, July 1977), vol. 2, exhibit 13. Estimates are costs for producing all steel products.
b. Speech by Edgar B. Speer, chairman of the board of the United States Steel Corporation, before civic leaders and government and agency officials, Conneaut, Ohio, February 23, 1977. Estimates are costs for producing flat-rolled products only.
c. Estimates are from confidential industry sources and are for flat-rolled products only.

near Conneaut, Ohio, on the shore of Lake Erie.[9] This plant was presumably to produce a limited array of flat-rolled products and would have a "new steel" capacity of 3 to 4 million tons. The company estimated that in 1976 dollars the new plant would cost $3 billion. Since Speer was apparently referring to finished steel capacity, the cost per net finished ton would be in the range of $750 to $1,000 per net finished ton. Assuming an 80 percent yield to finished products, the cost may be estimated at $600 to $800 a net raw ton in 1976 dollars, but some of this total may be for coal and iron mining or beneficiation facilities. It would appear that $600 a net raw ton represents a best estimate of the cost of all the Lake Erie steelmaking facilities in 1976 dollars. Escalated at the rate of increase of the nonfarm private business investment deflator, this translates into $689 per net raw ton in 1978.

One of the difficulties in calculating the costs of building new steel mills derives from the differences among the plants. Most recent plants use the BOF steelmaking process, but all do not have the same mix of finishing facilities. U.S. Steel's prospective Conneaut plant is envisioned to have only a plate and a hot strip mill. Burns Harbor has two plate mills, a hot strip mill,

9. Speech by Edgar B. Speer to civic leaders and government and agency officials, Conneaut, Ohio, February 23, 1977.

Table 4-2. *The Cost of Building a New Flat-Rolled Carbon Steel Plant per Million Tons of Raw Steel Annually*

Facility	Capacity (millions of tons)	Estimated costs (millions of dollars)		
		TBS (1975 dollars)	Industry (1978 dollars)	TBS[a] (1978 dollars)
Ore yard	0.66	5.65	. . .	6.75
Coal yard	0.78	6.68	25.00	7.98
Sinter strand	0.33	2.87	. . .	3.43
Scrap yard	0.38	2.08	. . .	2.49
Coke ovens	0.50[b]	61.50	222.50	73.49
Blast furnaces	0.95	43.97	. . .	52.54
Basic oxygen furnaces	1.18	29.44	35.00	35.18
Continuous caster	1.11	52.96	62.50	63.29
Hot strip mill	0.89	53.10	101.25	63.45
Pickling and oiling	0.84	27.02	. . .	32.29
Plate mill	0.14	23.26	55.40	27.80
Cold-reduction mill	0.50	95.00	62.20	113.53
Tinning line	0.14	25.40	n.a.	30.35
Galvanizing line	0.08	22.28	n.a.	26.62
Total without utilities	. . .	456.33	563.85	545.31
Utilities, spares, start-up	. . .	168.74[c]	208.55	201.64
Total	. . .	624.97	772.40	746.95

Sources: Temple, Barker, and Sloane, Inc., *Analysis of Economic Effects of Environmental Regulations on the Integrated Iron and Steel Industry* (U.S. Environmental Protection Agency, July 1977), vol. 2, exhibit 13; and confidential energy sources.

a. Estimate obtained by inflating 1975 TBS cost by growth in nonresidential investment deflator.
b. Author's estimate.
c. Assumes that these costs are 27 percent of total cost.
n.a. = Not available.

a cold-reduction mill, and a tinning line, but no galvanizing facilities. Neither has a bar mill, wire mills, or pipe and tubing facilities.

In order to simplify comparisons I have reconstructed the 1975 TBS data to reflect a flat-rolled carbon steel plant using 100 percent continuous casting. These data are shown for 1 million tons (raw steel) of capacity, assuming 90 percent capacity operation, in table 4-2. Tinplate, galvanized sheet, cold-rolled sheet, plate, and hot-rolled sheet are produced in the proportions shown. Utilities, spares, engineering, and start-up costs are assumed to be 27 percent of the total. These investment costs total $625 per raw ton in 1975 dollars. Escalated to 1978 dollars, using the deflator for nonresidential investment, yields $747 per raw ton. The 1978 industry estimate for the same facilities is $772 per raw ton, but it excludes tinplate and galvanized steel capacity. Granted that part of the latter estimate probably reflects investment in site

Table 4-3. *Estimated Capital Costs for U.S. New Steel Mills*
Current dollars

Year	*Cost per net raw ton*
1960	292
1961	290
1962	292
1963	295
1964	297
1965	301
1966	310
1967	320
1968	334
1969	350
1970	371
1971	395
1972	411
1973	428
1974	481
1975	558
1976	603
1977	708
1978	750

Sources: See text. Interpolations are based on the nonresidential investment plant and equipment deflator of the Bureau of Economic Analysis.

preparation, utilities, and other facilities that can be spread over future expansions, the best estimate for 1978 is undoubtedly less than $800 for the flat-rolled plant.

For later analysis I shall assume that the cost of building a new integrated mill with a relatively complete array of finishing facilities and all site preparation costs (at the time of breaking ground) is equal to the Burns Harbor cost in 1969, the ADL-TBS estimated cost in 1975, and $750 per net raw ton in 1978. All other yearly estimates are interpolations based on the Bureau of Economic Analysis (BEA) nonresidential plant and equipment deflator or on backward extrapolations from 1969 using the same deflator. These estimates are given in table 4-3.

International Comparisons of Capital Costs

It is never easy to compare capital costs among countries because of the difficulties in knowing precisely what share of the site preparation and other overhead capital costs is absorbed by the steel company and how much is provided by the government. Moreover, the array of facilities is likely to differ even more among countries than within the United States. Finally, as in the case of the United States, there are simply not many plants being built at any one time, and the most interesting data are not always public.

Despite these difficulties it is possible to assemble information from a variety of published sources, from steel engineers in various countries, and from the steel companies themselves. The compendium in table 4-4 reflects a combination of all these sources.

Two published studies of the 1976 cost of building comparable steelmaking facilities in Europe, Canada, Japan, and the United States reveal that the cost of building a complete new plant in Europe was 11 percent less than the cost in the United States and the cost in Japan was 28 percent less.[10] Jarvis finds even more striking cost differences, which Aylen argues are confirmed by other evidence on construction costs and purchasing power parities of the various currencies.[11] According to Jarvis the 1976 cost of building a plant in Europe was 22 percent below the U.S. cost, and the Japanese enjoyed a staggering 41 percent cost advantage in 1976. Adjusting the Barnett and Jarvis estimates for changes in the price indexes for capital formation and for changes in exchange rates narrows these differences by 1978. The Barnett estimates, when adjusted in this fashion, result in parity between the United States and Europe but a 10 percent cost advantage for Japan. Adjusting the Jarvis estimates in this fashion narrows the European cost advantage over the United States to 12 percent and the Japanese advantage to 27 percent.

Steel engineers and steel companies in Europe and the United States are generally of the opinion that U.S. and European capital construction costs are virtually the same but that eastern Asia offers the best location for building new plants. Other parts of the world, notably Latin America and the Middle East, may require costs 10 to 15 percent above U.S. new-plant costs. The estimates at the bottom of table 4-4 confirm this judgment about eastern Asia. Taiwan and Japan have costs that may be as much as 20 percent lower than U.S. new-plant costs, but countries such as Turkey or Indonesia appear to

10. D. F. Barnett, "Comparative Capital Costs in World Steel Industries," in D. F. Barnett, *The Canadian Steel Industry in a Competitive World Environment*, vol. 2: *Costs and Performance* (Ottawa: Resource Industries and Construction Branch, Industry, Trade and Commerce, 1977).

11. A. J. Jarvis, "Inflation and Capital Investment in the U.K.," *Transactions of the 5th International Cost Engineering Congress* (Utrecht: November 1978); and Jonathan Aylen, "Innovation, Plant Size, and Performance: A Comparison of the American, British, and German Steel Industries," paper presented at the Atlantic Economics Association Conference, Washington, D.C., October 12, 1979.

Table 4-4. *Estimates of the Cost of Building New Integrated Steel Plants in Various Countries*
Costs in dollars per net raw ton

Estimate source and region	1976 Slabs	1976 Finished steel	1976 Index (U.S. = 100)	1978[a] Slabs	1978[a] Finished steel	1978[a] Index (U.S. = 100)
Barnett[b]						
United States	399	726	100	468	853	100
Canada	400	756	104	403	762	89
Japan	332	525	72	482	764	90
European Coal and Steel Community[c]	368	649	89	488	860	101
Aylen[d]						
United States	...	...	100	...	...	100
Japan	...	...	78	...	...	88
Europe	...	...	59	...	...	73
Industry estimates[e]						
United States	...	...	...	...	900	100
United Kingdom	...	...	...	382	886	98
Canada	...	...	...	438	...	...
Australia	...	...	...	503	...	...
Indonesia (direct reduction)	...	...	...	...	900	100
Turkey (direct reduction)	...	...	...	...	862	96
Taiwan	...	...	...	...	651	72
Japan	...	...	...	...	721	80

a. Author's estimate, using domestic gross-capital formation deflators and changes in average exchange rates, 1976–78.
b. D. F. Barnett, "Comparative Capital Costs in World Steel Industries," in D. F. Barnett, *The Canadian Steel Industry in a Competitive World Environment*, vol. 2: *Costs and Performance* (Ottawa: Resource Industries and Construction Branch, Industry, Trade and Commerce, 1977).
c. Includes Belgium, France, Italy, Luxembourg, Netherlands, and West Germany.
d. Jonathan Aylen, "Innovation, Plant Size, and Performance: A Comparison of the American, British, and German Steel Industries," paper presented at the Atlantic Economics Association Conference, Washington, D.C., October 12, 1979.
e. From confidential industry sources.

suffer a slight cost disadvantage. In part the reason for high capital costs in other areas, such as the Middle East and South America, is the limited availability of local skilled labor. Since labor may account for 40 to 50 percent of all construction costs in a developed country, considerable savings are potentially available from using lower-wage labor in the less developed countries.[12] The necessary skills are often lacking; as a result, it is not unusual to find a project completed with less than 30 percent local value added. Given

12. U.S. industry confidential estimate.

the added costs of shipping equipment from the developed countries, the difficulty in using local labor for the skilled occupations required in plant construction ensures that the plant will cost as much as or more than one in the United States or in Europe.

An Operating Profile of Greenfield Facilities

Surprisingly, there are only minor differences in basic plant design among countries despite the substantial differences in the relative prices of labor and capital. The only significant variation from the standard blast furnace/BOF/ rolling-mill practice is the use of direct reduction of ore in conjunction with electric furnaces in areas where natural gas is available at 50 cents per million British thermal units or less.[13] Some of the plants now being completed in Venezuela, Indonesia, and the Middle East use a different technology up to liquid steel, but thereafter the forming and finishing processes are virtually identical to those in the rest of the world. As a result, cost differences are largely related to wage rates, capital charges on equipment, and transportation. In the analysis that follows, transportation costs for raw materials and products are ignored.

The data in table 4-2 suggest that a new greenfield site would probably require about $750 (in 1978 dollars) per net raw ton for a large integrated facility with an average complement of rolling mills for flat-rolled carbon products. Assuming that an 80 percent yield could be achieved from raw steel to finished product, in part through greater use of continuous casting, the cost per net finished ton would be $937.50 in 1978 dollars.

There are no published estimates of the cost of operating a new integrated carbon steel plant in the United States. Most of the cost savings over current plants would come from a reduction in labor utilization and lower energy consumption, due in large part to increases in the yield of final products from raw steel. The potential labor savings may be deduced from a comparison of Japanese practice with U.S. operating conditions. The average flat-rolled carbon steel plant required 8 man-hours per finished ton at 90 percent capacity utilization in 1978.[14] This is the experience of Inland Steel's only plant at Indiana Harbor.

13. Kaiser Steel Corporation engineers' estimate.

14. AISI reports that 8.95 man-hours a ton were required in 1978 to produce the average steel mill product. This includes alloy and stainless products as well as wire and tubular products. Therefore, the industry average for flat-rolled carbon steel products was probably 10 percent less, or about 8 man-hours a ton. (See American Iron and Steel Institute, *Steel at the Crossroads: The American Steel Industry in the 1980's* (Washington, D.C.: AISI, January 1980), p. 37.

Table 4-5. *Direct Man-Hours per Net Ton of Flat-Rolled Carbon Products, Large U.S. Plant, 1975*

Process or product	*Tons*[a]	*Man-hours per ton*	*Total man-hours*
Sinter strand	0.301	0.30	0.09
Ore yard	0.592	0.06	0.04
Coal yard	0.700	0.06	0.04
Scrap handling	0.280	0.20	0.06
Coke ovens	0.489	0.43	0.21
Blast furnace	0.859	0.29	0.25
Basic oxygen furnace	1.024	0.43	0.44
Continuous casting	0.105	1.05	0.11
Ingot pouring	0.895	0.03	0.03
Ingot breaking	0.746	0.25	0.19
Hot strip mill	0.685	0.42	0.29
Plate mill	0.105	2.75	0.29
Pickling and oiling	0.646	0.30	0.19
Cold-rolling mill	0.383	0.95	0.36
Galvanizing	0.063	0.98	0.06
Tin mill products	0.105	0.63	0.07
Total products	0.811	. . .	2.72
Addendum			
Total direct hours per ton: 2.72 ÷ 0.811 = 3.35.			

Source: Based on Temple, Barker, and Sloane, Inc., *Analysis of Economic Effects of Environmental Regulations on the Integrated Iron and Steel Industry* (U.S. Environmental Protection Agency, July 1977), vol. 2, exhibits 8A and 8B.

a. Material processed per ton of raw steel yield.

The best practice is probably closer to 7 man-hours a ton if the 1975 TBS study is accurate. Using only the man-hours for the "large" facilities and assuming a fairly typical product mix of hot-rolled, cold-rolled, galvanized, and tin-coated sheets, one can calculate the direct man-hours at 3.35 per finished ton. (See table 4-5). Given indirect man-hours of approximately the same magnitude and the labor required for packaging, shipping, and general administration, total man-hours a finished ton were undoubtedly somewhat more than 7 even for this largest model U.S. plant in 1975. By 1978 the best practice was probably about 7 man-hours for a flat-rolled carbon steel mix. The Burns Harbor plant, for example, might approximate this total if it had a full complement of tinplate and galvanizing capacity.

Even if a new flat-rolled carbon steel plant were to employ the best in-line design, it is difficult to see how it could produce a full mix of products in the United States with much less than 6 man-hours of labor, including all indirect and general, sales, and administrative labor. A comparison of direct

Table 4-6. *Comparison of Direct Man-Hours per Net Ton for Selected Processes and Products in Large U.S. and Japanese Mills*

Process or product	Man-hours per net ton: United States (1975)	Man-hours per net ton: Japan (1977)
Process		
Ore handling	0.06	0.05[a]
Coal handling	0.05	0.05[a]
Scrap handling	0.20	0.15[a]
Sinter strand	0.30	0.03
Coke ovens	0.43	0.25
Blast furnaces	0.29	0.10
Basic oxygen furnaces	0.43	0.26
Continous casting	0.40[a]	0.16
Ingot breaking	0.38	0.38
Hot strip mill	0.42	0.36
Cold-rolling mill	1.25[b]	1.18
Plate mill	2.75	1.75
Product		
Slabs	1.46	0.77
Hot bands	1.95	1.05
Cold-rolled sheet	3.61	2.09
Plate	4.58	2.61

Sources: Temple, Barker, and Sloane, Inc., *Analysis of Economic Effects of Environmental Regulations on the Integrated Iron and Steel Industry* (U.S. Environmental Protection Agency, July 1977), vol. 2, exhibits 8A and 8B; and Japan, Ministry of Labor, *Rodo Seisansei* (Tokyo, 1978).

a. Author's estimate.

b. Includes pickling.

man-hours per ton of the largest U.S. and Japanese mills in table 4-6 shows the U.S. industry to be using over 40 percent more direct labor for the average product. Since the U.S. estimate is for 1975 and the Japanese data are for 1977, this comparison undoubtedly overstates the difference somewhat. Savings in indirect labor and general, sales, and administrative costs are undoubtedly much more modest. Therefore, the total labor savings available to the U.S. industry if it could emulate the best Japanese practice might be 25 to 30 percent of current labor requirements. This would place the minimum estimate of labor requirements at about 5.5 man-hours a ton. A more realistic estimate, given U.S. labor institutions, is probably 6 man-hours a finished ton for the full mix of flat-rolled products.[15]

15. One frequently hears that Bethlehem's Burns Harbor mill is already more efficient than this would suggest, using between 4 and 5 man-hours a ton. This plant does not have a full complement of finishing facilities, however, nor does this estimate include the sales, general, and administrative man-hours required.

Table 4-7. *Current Average Practice versus Greenfield Plant Costs for a Flat-Rolled Carbon Steel Plant in the United States*
1978 dollars per net finished ton

		Greenfield cost	
Item	*Current practice*	*Crandall estimate*	*AISI estimate*[a]
Labor	118	88	66
Coal	50	40	32
Iron ore	50	50	48
Scrap	9	9	9
Miscellaneous	97	77	71
Capital	30[b]	161[c]	129
Total	354	425	355

a. Based on data in Arthur D. Little, Inc., *Steel and the Environment: A Cost Impact Analysis*, A report to the American Iron and Steel Institute (Cambridge, Mass.: Arthur D. Little, May 1975).
b. Annual outlays required to maintain plant at current efficiency.
c. Assuming 0.172 capital charge.

The other major source of operating cost savings derives from lower energy use. The most modern blast furnaces would allow U.S. firms to save about 20 percent in metallurgical coal use. Another $20 per net ton can be saved through reduced energy and miscellaneous savings, induced in large part by the increase in product yields of nearly 10 percentage points. The total cost savings over an average 1978 plant are thus $30 in labor costs and $30 in energy and miscellaneous costs per net finished ton, or $60 a ton.[16]

The estimate above of operating cost savings is very close to that claimed by Stelco for its new Nanticoke mill in Canada. While this mill is far from complete, Stelco's chairman, Peter Gordon, has said that it will generate operating cost savings of about 15 to 20 percent of Stelco's costs at its Hamilton, Ontario, plant.[17] My estimated $60 savings a net ton is 18.5 percent of 1978 operating costs of $324 a net ton (see table 4-7). The U.S. industry may be able to improve on Stelco's cost savings since Stelco's Hamilton plant is undoubtedly more efficient than the average U.S. plant. Assume that at the same input prices, the Hamilton plant has operating costs that are 10 percent lower than the average U.S. mill. Then a 15 to 20 percent improvement in Canadian costs would be equal to a 23.5 to 28 percent reduction in U.S. costs, rather than 18.5 percent. But this is the most optimistic set of assumptions; engineering analyses usually overstate the benefits of new technology.

16. This estimate is sensitive to the cost of energy. Were oil and coal prices to increase at a more rapid pace than the general rate of inflation, the cost savings would be greater. The labor cost savings are based on assumptions of $14.69 an hour and two man-hours less labor required.

17. *American Metal Market*, February 29, 1980.

The AISI claims that much greater operating cost savings are available from new facilities. It claims that a greenfield plant could be operated with 4.6 man-hours of labor, yielding a total cost saving of almost 30 percent.[18] Were such savings available, a greenfield plant would look much more attractive, as table 4-7 demonstrates.

The annual capital charges for a new facility may be computed as follows:[19]

The cost of capital is a mix of debt and equity costs. The cost of debt capital in 1978 may be approximated by the Aa corporate bond rate, or 8.7 percent. Recall from chapter 2 that the cost of equity capital for steel firms is equal to the average cost of equity for New York Stock Exchange firms. Assuming that the risk premium demanded by investors for the stock exchange composite is 6 percentage points above the average short-term yield,[20] that this average short-term yield was 5.8 percent over the past four years, and that the marginal tax rate for steel firms is 0.40, the cost of equity capital before taxes is 19.7 percent. With three-fourths equity and one-fourth debt, a steel firm would pay an average of 16.9 percent in 1978 for the capital required for a new plant. For a twenty-five-year life the capital charge on this plant would have to be 17.2 percent. Given a net cost of $937.50 a finished ton for the facility, the capital charges per ton would be 0.172 times $937.50 = $161.25.

The capital charges for a new mill are staggering when compared with the cost of maintaining old equipment. In the past decade the steel industry has been spending $25 (in 1978 dollars) a finished ton of capacity to maintain and refurbish its plants, including the cost of pollution control. Assuming that $30 a net finished ton is sufficient to maintain viable facilities in their current condition, the incremental capital requirements per ton are $131 less than those required by a new plant. Obviously the operating cost savings over an average plant do not offset this monumental difference, regardless of one's assumptions about energy and labor usage.

The comparison of unit costs at full practical capacity for a flat-rolled carbon steel plant in table 4-7 shows that "modern" new plants have higher costs than does the average existing plant. Existing plants can produce the average product for $324 a ton plus $30 to maintain the capital facilities at constant productive capability. An efficient plant, such as Burns Harbor or Inland's Indiana Harbor mill, could improve on this performance by at least

18. Author's discussion with AISI members, April 23, 1980.

19. See chapter 6 for greater detail.

20. Merton has recently argued that this risk premium is greater than 6 percent. See Robert C. Merton, "On Estimating the Expected Return on the Market: An Exploratory Investigation," Working Paper 444 (National Bureau of Economic Research, February 1980).

$15 a ton and perhaps more. A new plant would reduce operating costs by $60 a ton, according to my estimate, but capital costs would rise by $131 a ton. Therefore, unit costs would rise to $423 a ton. The AISI estimate of cost savings (shown in the last column of table 4-7) is more substantial, reflecting a reduction of labor costs of 44 percent and energy costs of 36 percent. The AISI thus projects operating cost savings of about 30 percent, substantially more than the recent Canadian experience with the same technology.

In 1978 U.S. carbon steel prices averaged nearly $385 a net ton. Excluding wire, tubular, and bar products would lower this estimate to $360 a net ton. Capacity expansion at those prices might be justified only if anticipated inflation results in widening gross margins. For instance, the present value of a new plant, *PV,* with a twenty-five-year life is given by

$$PV = \int_0^{25} (R_o - C_o)\, e^{(g-r)t}\, dt$$

where $R_o - C_o$ is the gross margin currently obtainable, g is the expected growth in operating costs and prices, and r is the opportunity cost of capital before taxes. Assume that $R_o - C_o$ per ton for a new plant (from table 4-7) is $360 − $264 = $96, r = 0.172, and the anticipated inflation rate over the twenty-five-year life of the plant is 6 percent. The present value per ton for this plant is $805 if inflation is balanced at 6 percent for the life of the plant and if the plant operates at practical capacity for its entire twenty-five-year life. A more reasonable assumption would be that the plant would not be able to enjoy 90 percent capacity production ("capability") for this twenty-five-year period. If, for example, the plant could only achieve an average of 95 percent of this assumed capability, its value would fall to $765 a ton. In either case the plant is worth substantially less than its construction costs of $937.50 a ton. It could only be justified, under my assumptions, if the 1978 capital markets were discounting twenty-five-year inflation rates of substantially more than 6 percent a year.

As a final check on these results, one can compare current market valuations of assets with estimates of the cost of new steel plants. For a full array of steel products, the cost of building a new plant is probably somewhat more than $1,000 per net finished ton. Current assets are valued at $70 per net finished ton. At a 0.172 capital charge and zero nominal growth in cash flow, the market is estimating current cash flow at $12 per net ton, or very close to the estimate based on table 4-7. Cash flow would have to rise to more than $172 per net ton, an increase of $160, for a new plant to be viable. If $30 in annual refurbishing costs can be saved, this means that a new plant must have $130 less in operating costs than the average current plant. Since an average plant has operating costs of $350 or more per net ton, the operating cost savings must be more than 35 percent.[21] Thus the savings must be substantially more than those suggested by the recent Canadian experience—an unlikely possibility.

Modernization of U.S. Mills

A number of industry spokesmen argue that modernizing existing facilities dominates the greenfield strategy. Indeed, this is the strategy that Inland is following at Indiana Harbor and that the AISI now advocates.[22] Unfortunately the economics of modernization cannot be characterized in a general way. The additional capital required may be as low as $500 per net raw ton or as high as $800 depending on the configuration of the plant. The Inland expansion is apparently at the high end of this range,[23] but the operating costs of a fully rounded-out facility are not easily measured, nor are they likely to be similar across plants. Expansions could undoubtedly occur at Bethlehem's Burns Harbor facility, at U.S. Steel's Fairless plant, or at National Steel's Granite City or Great Lakes plants, but the scope of these expansions is necessarily limited by space and the design of the existing plant.

The AISI has analyzed the economics of a plant modernization program that would increase capital expenditures from $2.2 billion (1978 dollars) a year to $4.9 billion a year.[24] This program would expand capacity by 7.6 million tons of finished products over ten years and approximately double the capital spent on modernization. The payoff to this accelerated rate of modernization is based on much the same assumptions as the AISI greenfield analysis. Labor savings of more than 40 percent and energy savings of 35 percent on a full complement of modernized facilities are required to make the investments in a full array of new facilities appear profitable, assuming that the Capital Cost Recovery Act (House bill 4646) is passed. This legislation, which was not enacted by the Ninety-sixth Congress, would shorten the useful life for depreciation calculations to ten years for structures and five years for equipment. But even if this tax reform were enacted, the expanded modernization program would not be financially attractive at labor- and

21. These estimates are higher than those in table 4-7 because they include all steel products, not simply the flat-rolled carbon steel mix.

22. See AISI, *Steel at the Crossroads,* pp. 33–38.

23. The total cost of Inland's expansion at Indiana Harbor is expected to be about $780 per net raw ton of additional capacity (company estimate).

24. AISI, *Steel at the Crossroads,* pp. 21, 38.

energy-cost savings in the 20 to 25 percent range—a range that appears more likely.

It is impossible to assay the modernization possibilities in the U.S. industry without detailed information on each plant. It is clear, however, that despite large expenditures on capital equipment, the U.S. industry's profitability has not improved. In fact, the returns from these investments have been very poor, as the following crude analysis demonstrates:

In 1969 the firms reporting financial data to the AISI (representing approximately 90 percent of the industry) reported a total book value of stockholders' equity of $12,836 million.[25] The market value of this equity was equal to 0.7 times book value,[26] or $8,985 million, which translates into $15,993 million in 1978 dollars using the consumer price index to adjust for inflation. In 1978 the reporting companies (virtually the same group of firms) reported $18,403 million in equity, which had a market value of $8,097 million. In this interim less than $450 million in new equity had been issued. Thus over a ten-year period the reporting firms lost more than $8 billion (1978 dollars) in the market value of their outstanding equities. Over this same period these firms reported $24,539 million (1978 dollars) in reinvested cash flows and paid $7,161 million (1978 dollars) in dividends. In short, the steel firms plowed back almost $25 billion in earnings plus depreciation flows in pursuit of an uncompounded sum of $7 billion in dividends and a net reduction of $8 billion in the value of their stockholders' investment. The return on this $25 billion was undoubtedly negative for the 1969–78 period.

Unless the new modernization program envisioned by the industry is to be designed differently—particularly by being targeted on far fewer plants—it is difficult to see how an acceleration of investment in this industry can be profitable. Presumably the industry undertakes the most profitable investments first, rationing out the less profitable projects. Increasing investment flows must activate some of the latter, less profitable projects. It would be very difficult to argue that the *average* return on capital would rise with increased investment.

International Comparisons of Greenfield Operating Costs

While there are some estimates of the capital costs of building new facilities in various countries, the economics of operating new integrated steel facilities in areas where expansion is likely are not well known because experience is

25. All data are drawn from AISI, *Annual Statistical Report, 1978,* table 1B, p. 8.
26. Standard and Poor's, *Analyst's Handbook, 1979,* p. 73.

so limited. Since I am interested in estimating the costs of operating efficient new plants, I examine only those facilities that are likely to be justified on the basis of their economics alone. Partially integrated facilities in labor-poor Saudi Arabia or a landlocked plant in Turkey, designed to stimulate employment in a remote eastern region, are not included. Rather, large integrated facilities at coastal sites in South America or Eastern Asia would seem to offer the most interesting prospects.

Operating costs of a new steel plant are simply labor and materials (including energy) costs. Assuming that each plant is located at a coastal site, the cost of basic materials per ton of output should vary only in a minor way. Transportation costs to coastal sites for oil, coal, iron ore, or scrap do not vary by much across different parts of the world simply because bulk freight rates are very low and will not vary appreciably with modest differences in distance from the major iron- and coal-exporting areas of the world. Therefore, the major source of operating-cost differentials across different regions is labor cost.[27]

Obviously wage rates in South America and Asia are substantially below those in the United States, but the differences in average wage levels understate the differences in hourly labor costs for the steel industry. The U.S. iron and steel industry paid its labor 69 percent more than the average manufacturing wage (table 4-8).[28] In most other countries the premium is in the vicinity of 25 percent over the average. As a result, the U.S. industry suffers a competitive disadvantage that is much greater than general labor scarcity would suggest.

Hourly compensation for the steel industry in various countries is presented in table 4-8. The Bureau of Labor Statistics estimate for the United States is lower than the AISI estimate because the latter does not include all the smaller firms and the former excludes nonproduction workers. The difference between the Brazilian or the Korean labor rate and the U.S. rate is very large. It would be reasonable to assume that hourly labor costs for steel would be in the $2.00 to $2.50 range for Latin America in 1978 and $1.00 to $1.25 in eastern Asia (other than Japan). To be conservative, I use $3.50 and $2.00, respectively, for 1978. This allows for the upward pressure on steel earnings that would develop if major new investments were to occur in those areas. Given present labor market conditions, this obviously overstates the wage bill in these less developed countries.

27. Some differences in energy usage may result from the variance in product yields that result from differences in management and labor efficiency. (I am indebted to Kiyoshi Kawahito of Middle Tennessee State University for this point.)

28. This is slightly lower than the estimated premium in table 2-8 because the latter is based on data from AISI-member companies, and AISI does not include all the iron and steel industry firms covered in the Bureau of Labor Statistics data.

Table 4-8. *A Comparison of Steel Industry and Average Manufacturing Wages, Selected Countries, 1978*

Country	Total hourly compensation for production workers (dollars per hour)		Compensation ratio of iron and steel to all manufacturing
	All manufacturing	*Iron and steel*	
United States	8.33	14.04	1.69
Europe			
Germany	9.48	10.60	1.12
Netherlands	9.77	12.01	1.23
Italy	6.18	6.98	1.13
Latin America			
Brazil	1.67	2.12	1.27
Venezuela	2.53	n.a.	...
Mexico	2.00	n.a.	...
Asia			
Japan	5.47	8.22	1.50
Korea	0.85	n.a.	...
Taiwan	0.80	n.a.	...
Hong Kong	1.13	n.a.	...

Source: Unpublished data from the Bureau of Labor Statistics.

It is probably unrealistic to assume that greenfield plants in Asia (outside Japan) or South America will achieve a productivity level equal to that in the United States or Japan for similar facilities. For the purpose of this analysis I assume that labor productivity is directly related to the wage rate and that 10 man-hours a ton are required to produce finished products from a diversified, fully integrated mill in eastern Asia (other than Japan) and Latin America. These productivity estimates compare with 6 to 7 man-hours a ton for the United States or Japan in modern facilities.

For capital costs I assume that construction costs are $1,000 per net finished ton in South America (approximately 107 percent of U.S. costs) and $800 in eastern Asia (approximately 85 percent of U.S. costs). A 20 percent capital charge—a 2.8 percent risk premium over U.S. rates—is used to calculate annual capital charges. Materials costs are assumed to be $176 per net finished ton, the same as in Japan or in the United States.

The full comparison of Asian and Latin American expansion with the United States is presented in table 4-9. The Asian location shows an enormous advantage of $71 a ton over the U.S. mill and very nearly competes with best U.S. practice at existing plants, ignoring a return on equity in the latter. The South American facility is only marginally better than the new U.S. plant,

Table 4-9. *Comparison of New-Plant Unit Costs in the United States, Asia, and South America (per Net Finished Ton of Flat-Rolled Carbon Steel)*
Costs in 1978 dollars

Costs and assumptions	*United States*	*Eastern Asia (excluding Japan)*	*South America*
Labor	88	20	35
Raw materials and miscellaneous expenses	176	176	176
Capital charges	161	160	200
Total	425	354	411
Assumptions			
Man-hours per ton	6	10	10
Wage rate per hour	14.69	2.00	3.50
Construction cost per ton	937.50	800.00	1,000
Capital charge	0.172	0.20	0.20

Source: Author's calculations.

perhaps 2 percent lower in full costs per ton, assuming lower productivity and higher capital costs than in the United States.

This analysis is very conservative in its assumptions concerning the potential cost advantages of Asian or Latin American locations. The capital-cost and labor-productivity disadvantages projected for both areas may be overstated and of short duration. Surely the competition among Japanese, European, and North American steel engineers to sell and install technology in these developing areas must reduce these disadvantages. In many countries social institutions may militate against the successful construction of a steel industry, but in others, such as Korea, Brazil, Taiwan, or Venezuela, there is no reason to believe that efficient new mills cannot be built. Nor is there reason to believe that their wage levels will rise to the levels enjoyed by U.S. steelworkers in the future. This does not mean that these emerging economies will begin to dominate the world's steel industry, but it is possible that capacity expansion in these countries may keep the real price of steel very close to its current level.

Conclusion

Given the labor-intensity in the production of steel and in the construction of steelmaking facilities, the United States is not in a favorable position to expand steel capacity through the construction of new integrated works.

Eastern Asia and South America are likely to attract new steel investment. Korea, China, and Taiwan have low-cost labor, coastal sites, and apparently low plant construction costs.[29] These countries can operate new facilities at costs that are probably at least $70 lower per net finished ton than operating costs in prospective new U.S. mills.

The best *existing* U.S. mills have operating and incremental capital costs that are very close to the costs of operating and amortizing the capital investment of *new* mills in eastern Asia. Therefore, while new mills are not likely to be built in the United States in the foreseeable future, at current exchange rates the most efficient U.S. mills should have little difficulty defending their home markets from major increases in import penetration from even the most efficient exporting countries. The U.S. firms' share of the home market may decline, but it will be a slow decline as the less efficient U.S. mills close. Similar conclusions hold for Canadian steel companies that now enjoy an advantage over their U.S. rivals because of the depreciation of the Canadian dollar. Indeed, incremental expansion by Stelco and Dofasco, Canada's two largest firms, may pose a greater threat to U.S. firms located in the Great Lakes region than the newest of mills in Latin America or Asia.

29. Some argue that these advantages are offset in part by the requirements for building infrastructure, but there is no documented evidence that this infrastructure is more expensive to build in these countries than in the developed world.

V

U.S. Trade Policy for Steel

Given the importance of steel in any industrial economy, it is not surprising that the steel industry receives considerable attention in the formulation of international trade policies. For advanced economies this importance is magnified by the inevitable decline in the viability of a relatively labor-intensive steel industry.[1] Thus the problems faced by U.S. steelmakers in adjusting to foreign competition are similar to those experienced by German, French, Belgian, or British steel producers. And there are even signs that the Japanese edge in steel production may be dulling somewhat as the wage rate in Japan rises with the value of the yen.

Presumably the objectives of U.S. trade policy are to foster a climate in which the steel industry can adjust to changing market forces with limited disruption for the owners of capital and labor. The most recent statement of these objectives is to be found in the Solomon Report of December 6, 1977, announcing the "trigger" price system:

> Our primary objective is to assist the steel industry in a manner which will stimulate efficiency and enable the industry to compete fairly.
>
> A second objective is to help ease the burden of adjustment to market trends for both industry and labor.[2]

Nowhere in the Solomon Report is it asserted that the United States needs a steel industry that is larger than a free world market would yield. The industry is only to be helped in its struggle with foreign competitors by assuring "fair" competition and reasonable domestic tax and environmental policies.

The Solomon Report was issued almost two decades after the U.S. industry began to exhibit its relative decline in world markets. Why was the "burden of adjustment" still given as a reason for trade protection when most of the adjustment occurred in the 1960s? There are two answers to this question: (1) three plants closed in 1977; and (2) the administration was reacting to

1. Even the modern Japanese industry is not very capital-intensive. Labor's share in value added in the Japanese steel industry is very close to the average for all manufacturing.

2. Solomon Report, pp. 7–8.

congressional pressures to ease the burden of price competition, which had grown fierce in the deepest world recession since World War II.

Arguments for Trade Protection

Magee summarizes the prospective arguments for trade protection succinctly.[3] They include (1) changing the domestic distribution of income, (2) increasing aggregate employment, (3) improving the balance of payments, (4) exploiting monopoly and monopsony power in world markets through an "optimum tariff," (5) protecting infant industries, and (6) promoting the national defense. He essentially dismisses the first four reasons because more efficient policies are available to accomplish these goals and because the use of trade protection to accomplish them generally requires that other countries not retaliate, an unlikely assumption. Of the remaining two, only the promotion of national defense is likely to have any merit as a rationale for a protectionist steel policy. Clearly the U.S. steel industry is not an "infant" industry. Arguments for trade protection, therefore, do not center on creating a transition to maturity. Instead, they are a curious blend of protection from "unfair" competition and "national security." The former argument is basically one of cyclical protection from employment disruptions caused by dumping; the latter is rarely presented in any formal manner.

Unfair Competition

As chapter 3 indicated, the decline of the U.S. industry since 1960 can be explained on the basis of changing relative input prices in an increasingly competitive world market. Nevertheless, it is possible that unfair competition has added to this decline, accelerated it, or simply made cyclical instability much worse. But what is "unfair" competition?

Dumping. The Trade Act of 1974 amended the Antidumping Law to expand the definition of dumping.[4] Dumping was historically defined as sales abroad at prices below the net realizations in the home market. This dumping might or might not be below average (or even incremental) cost. The Trade Act extended the definition of dumping to include substantial sales below the

3. Stephen Magee, "The Welfare Effects of Restrictions on U.S. Trade," *Brookings Papers on Economic Activity, 3:1972*, pp. 650–55.

4. The major change in the Antidumping Law of 1921 is contained in section 205(b), which now requires that all sales to the United States be above the cost of production if there is domestic injury from such sales. Sales below the cost of production cannot be used to determine fair market value in such cases.

average cost of production over an extended period of time even when exports to the United States were not below the exporter's domestic price. If such sales constitute injury to the U.S. industry or its employees (per the International Trade Commission), the Treasury Department (or since 1979 the Commerce Department) may find that sales below the cost of production constitute dumping and require that exporters use "constructed value" in establishing the fair value of their exports. Constructed value is unit costs plus 8 percent for a profit margin with the requirement that overhead costs be equal to at least 10 percent of direct cost.[5] Thus during a recession in a competitive world market U.S. law may now require exporters to raise prices if they wish to sell to the United States at all since unit costs rise with declining volume. How are unit costs to be measured? The Trade Act is silent on this question. Indeed, for any reasonably complex production process, unit costs probably could not be measured with any precision even if a definition of these costs were provided.

Given the competitiveness of the world steel market and the substantial fixed costs in the industry, steel prices may fall substantially in slumps. The 1965–68 period witnessed substantial price cutting in European markets. In 1975 the deepest recession since World War II saw the average export price of steel fall by as much as 50 percent for some products, such as heavy plate.

Obviously 1975 was a year in which many producers in the world were operating at negative profit rates, but widespread charges of dumping in the U.S. market did not arise because imports actually *fell*. As the 1976 recovery aborted, imports into the United States began to rise, and by 1977 they had risen to record levels. This sharp rise in imports, a depreciating dollar, and the administration's encouragement led the U.S. companies to invoke the cost-of-production standard for the first time in their 1977 dumping complaints filed with the Treasury Department.[6]

The alleged unfairness of cyclically sensitive pricing is often based on the assertion that it could not occur if exporters were not able to engage in price discrimination, charging higher prices in domestic markets and selling at prices as low as marginal costs abroad.[7] There is an element of truth to such

5. Solomon Report, p. 11.

6. Cases were brought in 1977 against Japan, the United Kingdom, and five other European Economic Community countries by United States Steel Corporation, Armco, and National Steel Corporation, respectively.

7. The most lucid case against this form of pricing is found in Putnam, Hayes and Bartlett, Inc., *The Economic Implications of Foreign Steel Pricing Practices in the U.S. Market* (Newton, Mass.: Putnam, Hayes and Bartlett for the American Iron and Steel Institute, August 1978). The case for allowing dumping is made in Robert W. Crandall, "Competition and 'Dumping' in the U.S. Steel Market," *Challenge*, vol. 21 (July-August 1978), pp. 17–18.

Figure 5-1. *Japanese and European Coal and Steel Community Export Prices versus U.S. Domestic Price for Cold-Reduced Sheet, 1967–76*

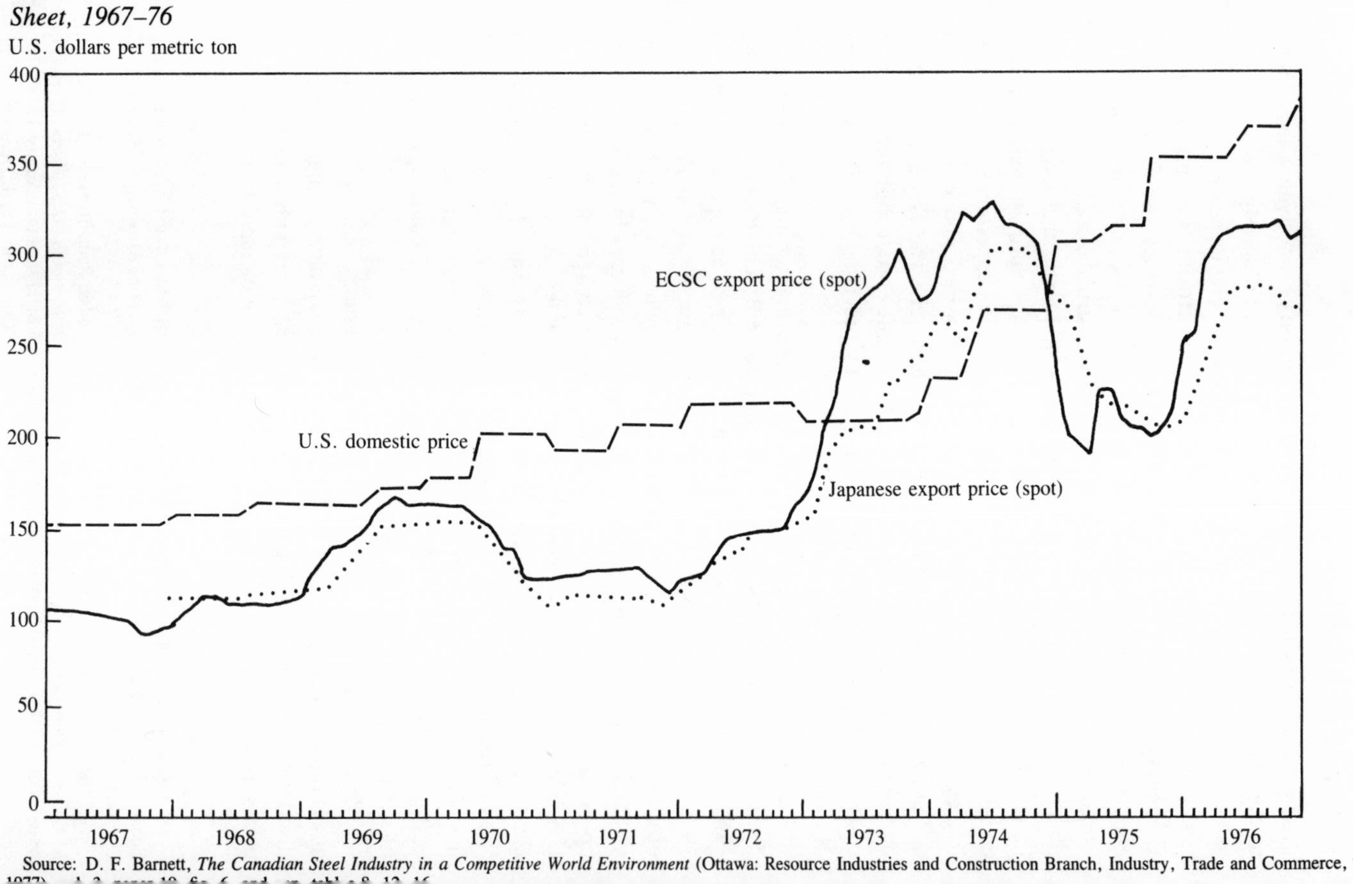

Source: D. F. Barnett, *The Canadian Steel Industry in a Competitive World Environment* (Ottawa: Resource Industries and Construction Branch, Industry, Trade and Commerce, —

an observation, for many steel enterprises (particularly in developing countries) might not exist were it not for the domestic protection afforded by their governments. But once in existence, their export policies are largely driven by competitive forces, as chapter 3 demonstrated, at least for the Japanese industry.

The reason for engaging in price competition in world markets is that it cannot be avoided in the struggle for sales. Nor does this strategy necessarily result in the subsidization of employment if steel producers in other countries compete on price. U.S. producers may lose market share at home, but only if they fail to engage in similarly competitive pricing. For example, it is not surprising that U.S. producers lost market share to importers in 1976 and 1977 when they failed to cut prices to meet the price competition from Japanese and European exporters in late 1974 and throughout 1975. No clearer picture of this difference in cyclical flexibility is available than that shown in figure 5-1. The published U.S. price for cold-rolled sheet moved upward throughout 1970–76 with only a few deflections, while European and Japanese export prices declined in both the 1969–70 and 1974–75 recessions. Obviously the failure to match this flexibility in pricing must have led to variations in U.S. producers' domestic market share.

There is neither an efficiency nor an equity argument against allowing the free importation of steel from countries that permit their steel industries to practice price discrimination. The fact that, say, Japanese consumers may on occasion pay more for steel products because of protectionism afforded steel or steel-fabricating industries in Japan has no direct bearing on the desirability of allowing Japanese steel firms to compete freely for U.S. sales. U.S. steel producers could not compete in Japan for carbon steel customers even under the most liberal of trade policies, and the price at which Japanese firms offer steel in the export market is not affected by home-market competitive conditions. In fact, the efficient policy toward monopoly, if the monopoly power cannot be dissolved, is to allow price discrimination. As long as the incremental sale in each market is made at a price equal to marginal cost, the conditions for efficient allocation are satisfied. Therefore, the argument that price discriminating monopoly undermines the presumption for liberal import policies has no basis in economic theory.

Subsidies. Many U.S. businessmen believe that different institutional arrangements used by other countries to channel capital resources into productive activities constitute subsidies and, therefore, unfair competition. The Japanese practice of having consortia of banks provide long-term financing through short-term loans that are rolled over periodically is deemed to reflect a subsidy because U.S. banking institutions would not extend such credit to firms in cyclical industries that had only 10 or 20 percent equity in their capital structures. Moreover, it is argued that nationalized industries are subsidized by definition if they do not rely on private capital markets for financing their investments.

These differences in capitalizing productive enterprises across countries are difficult to measure. For instance, it is not a simple matter to determine if steel producers were funded by Japanese banks on a more favorable basis than, say, textile or automobile manufacturers. Ex post, the investments appear to have been well made, but were they sufficiently attractive ex ante to justify such a large societal commitment to steel? Nor can one say with any assurance that Rumanian or Polish steel enterprises are "subsidized" without knowing the prospective returns from a variety of investments (including steel) in these countries.

The most direct and obvious forms of subsidy available to firms in market economies are those obtained through tax deductions, credits, or rebates. For instance, rebating excise or income taxes would be a subsidy to a particular industry if limited to that industry, but a general tax rebate to all industries (such as the rebating of value-added taxes on exports in Europe) is not a subsidy in the usual sense. Moreover, rarely are subsidies conferred without a quid pro quo.

In a recent report the Federal Trade Commission has estimated that a variety of financial subsidies to German steel producers was more than offset by the requirement that these steel firms buy expensive German coking coal.[8] The same report found that subsidies were negligible in all major steel industries except for the United Kingdom, where subsidies totaled about 3 to 4 percent of the value of steel output.[9] This study concluded that the Japanese system of channeling capital did not constitute a major subsidy for steel producers in that country, nor did the rebating of value-added taxes.

National Defense

Even if a protective trade policy is not required by the existence of foreign subsidies or other unfair trade practices, a protected steel industry may be a requisite of national security. Relying on foreign producers of steel or of any

8. U.S. Federal Trade Commission, *The United States Steel Industry and Its International Rivals: Trends and Factors Determining International Competitiveness* (Government Printing Office, 1978), p. 337.

9. Ibid., pp. 350–63.

other material in peacetime could create severe problems during a national emergency. For this reason strategic stockpiles have been maintained for various metals such as tin and alumina, which are not produced in large quantities in the United States. But steel is not one of these materials for a number of reasons.

First, steel capacity remains large in the United States. Even during peak demand, the United States continues to supply more than 80 percent of its own steel.

Second, steel imports come from a very large number of countries with quite different political systems and geographical locations. To suppose that a political coalition of, say, Canada, Brazil, Japan, Korea, Germany, Italy, Poland, South Africa, and India could form a cartel to deny the United States its steel requirements during a national emergency is not very realistic. Moreover, the distribution of world steel capacity is growing more dispersed with time, as figure 5-2 shows. Given the strong probability that future capacity growth will occur in countries other than the current major producers, as chapter 4 demonstrated, this dispersion can only grow.

Third, steel consumption for domestic uses can be postponed for a considerable period of time. The quantity of sheet steel used in automobiles and home appliances, for example, can easily be reduced during a period of crisis. Similarly the steel used in containers for consumer goods can be reduced for a substantial period. These combined uses total nearly 30 percent of all steel consumption, or 30 million tons in an average year (table 5-1). Even during the peak year of the Vietnam War, steel used for ordnance amounted to only 2.2 percent of total shipments.[10]

During World War II total U.S. mill shipments to industries identified with the war effort—construction, automotive, aircraft, and ordnance—were about 40 percent of the industry output. This is a somewhat arbitrary calculation since it might be argued that all output was directly or indirectly supporting defense in those years, but the critical shipments are surely represented in table 5-1. By 1976–77, a very similar percentage of U.S. shipments was destined for industries producing consumer goods, most of which represent postponable purchases for consumers in a time of war. There is no similar evidence on the distribution of imported steel across markets, but U.S. shipments averaged more than 85 percent of U.S. supply in 1976–77. Therefore, it would appear that at least 38 percent of U.S. steel supplies in peacetime are destined for consumer purchases that could be postponed for a military

10. *Annual Statistical Report: American Iron and Steel Institute, 1975* (Washington, D.C.: AISI, 1976), table 16.

Figure 5-2. *Lorenz Curves for Non-Communist World Steel Production, 1967 and 1978*

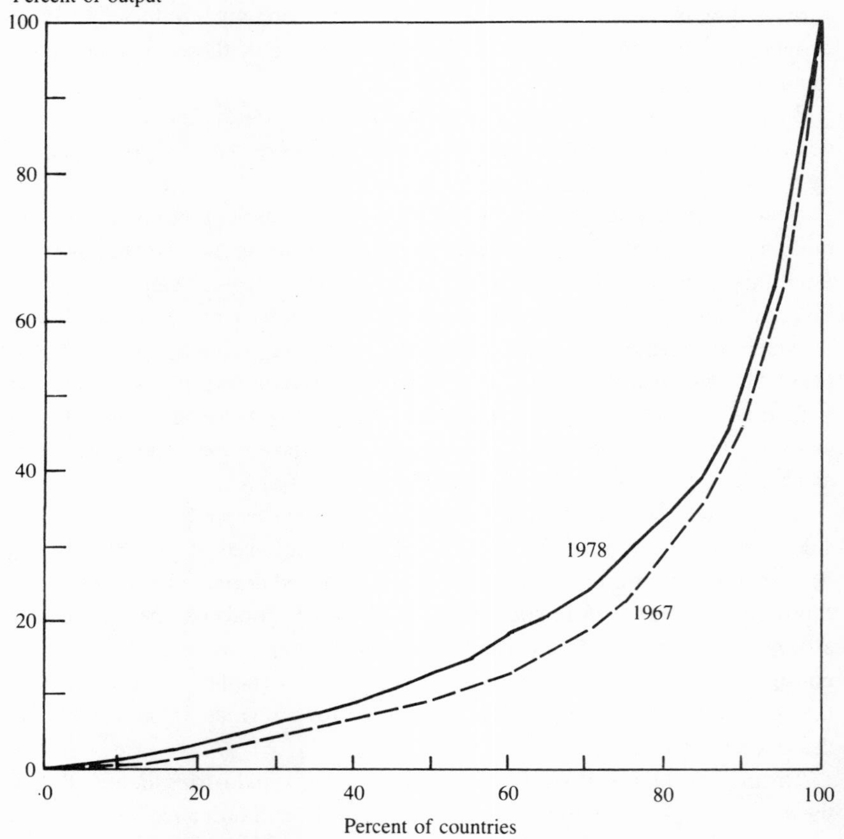

Source: *Annual Statistical Report: American Iron and Steel Institute*, 1967 and 1978 editions.

effort. This is enough to sustain a mobilization of the magnitude of 1942–44, the greatest national defense effort in U.S. history.

Finally, if steel supply during a protracted emergency were a major consideration in policy planning, inventories of semifinished steel could easily be increased. Rolling-mill capacity generally exceeds raw steel capacity; hence excess inventories of semifinished steel could be drawn down and rolled during an emergency. Of course, steel oxidizes in storage, and greater scarfing of stored semifinished steel would be required, but this additional cost would be small compared with the cost of maintaining additional capacity.

Table 5-1. *Distribution of U.S. Steel Shipments in Peace and War*
Thousands of net tons of finished steel

World War II			
Use	*1942*	*1943*	*1944*
Forging (auto and aircraft)	498	649	597
Construction			
Aircraft	639	205	97
Shipbuilding	666	220	209
Barracks and bases	2,443	1,422	1,484
Shipbuilding	9,440	11,509	10,287
Automotive stamping and forming	1,070	1,223	1,286
Automotive	1,601	1,631	1,507
Aircraft	521	887	532
Ordance, etc.	6,111	7,480	6,615
Total military-related	22,989	25,226	22,614
Total U.S. shipments	60,465	59,906	64,193
Addendum			
Percent of shipments for military uses	38.0	42.1	35.2

Peacetime		
Use	*1976*	*1977*
Automotive	21,490	21,351
Appliances and utensils	2,129	1,950
Other domestic and commercial equipment	1,846	1,813
Cans and closures	5,173	5,301
Total	30,638	30,415
Domestic mill shipments to direct users	68,109	67,065
Addendum		
Percent of shipments for consumer uses	45.0	45.4

Source: *Annual Statistical Report: American Iron and Steel Institute, 1946* (New York: AISI, 1947); and ibid., *1978* (AISI, 1979).

National Defense and the Optimal Capital Stock

Thompson has advanced two further national defense arguments that merit consideration.[11] The first assumes that national defense requires a large capital stock to serve as a deterrent to would-be aggressors. Investment thus carries an external economy in terms of increased value of the stock of national defense. Efficient allocation of resources requires that investment be subsidized so as to equate the social rates of transformation and substitution between consumption flows in different time periods. This argues for a general subsidy to basic industry, however, not a specific subsidy to steel or cement.

The second argument advanced by Thompson also invokes economic efficiency as a reason for national defense subsidies. Investors anticipate that periods of war and other emergencies will generate pressures for price controls on the output of investment goods. As a result, they invest less in these goods than is required for intertemporal optimality. Were the threat of price controls removed, there would be no reason for subsidizing investment goods in critical

11. Earl A. Thompson, "An Economic Basis for the 'National Defense Argument' for Aiding Certain Industries," *Journal of Political Economy*, vol. 87 (February 1979), pp. 1–36.

industries. But with a real threat of price controls, some offsetting public policy is required. This may take the form of import protection of domestic capital subsidies.

Steel is an obvious candidate for formal or informal price controls during any period of war or even in peacetime inflation. Steelmakers clearly do not feel that they can price freely during booms in peacetime or during war. Informal "jawboning" has been a part of government policy since the Truman administration, and it must affect investment decisions if it has any impact on prices.

If one accepts the theory that actual capital investment is below socially optimal investment in the steel industry because of the threat of wartime price controls, the amount of the countervailing subsidy required to obtain the optimal capital stock depends on the degree to which prices are controlled during the war and the probability of a protracted war. Thompson points out that sellers of iron and steel are likely to be able to avoid price controls to some extent in sales to civilian industries. Purchases by the government are likely to be made at less than otherwise market-clearing prices, but direct sales of iron and steel to the military represent a relatively small share of steel sales. In the peak World War II years between 35 and 42 percent of steel mill products were shipped to industries largely engaged in defense-related activities. Of these sales, however, only 30 percent was ordnance sales, less than 10 percent was for construction of barracks and bases, while the remainder was to the automobile, aircraft, shipbuilding, and other defense-related industries. It is difficult to say how effective government price controls were in keeping sales to these latter industries at less than market-clearing prices.

Thompson argues that the probability of a protracted (four-year) war involving the United States is 1:50. If, during such a war, steel prices were held on the average to a level generating only one-half of the cash flow to investment that would derive from market-clearing prices for four consecutive years and if a steel investment lasts twenty years, generating 5 cents in present value in each year for each dollar invested, the "national defense" external social value from the investment is

$$\frac{1}{50}\{[(0.05 + 0.05 + 0.05 + 0.05) \times 17] + (3 \times 0.05) + (2 \times 0.05) + 0.05\} = \frac{3.7}{50} = 7.4 \text{ percent.}$$

A subsidy of 7.4 percent would be required even if prices were so low as to reduce cash flow by 50 percent of market-clearing values during the four-year war. This is obviously substantially below the subsidy required to produce investment in new steel plants in the United States. (See chapter 4.)

Whatever the reason, the U.S. steel industry has enjoyed some form of trade protection for most of the past ten years in addition to the traditional tariff protection. This trade policy has taken the form of "voluntary" quantitative restrictions of European and Japanese exports (1969–74) and trigger prices (1978–80). The next section deals with the effects of these policies.

The Effect of Trade Protection for Steel since 1969

Imports of steel products into the United States were negligible before 1960. Despite this fact, tariffs were far from inconsequential, averaging 6 to 8 percent of the value of imports. Until the late 1960s, however, nontariff barriers were not an important influence on imports into the United States. Since 1968 two sets of nontariff barriers have limited imports: (1) the Voluntary Restraint Agreements of 1969–74 and (2) the trigger-price system erected in 1978 as a temporary method of ensuring fair competition. In this section the impacts of each of these import policies are reviewed.

The 1969–74 Voluntary Restraint Agreements

In response to the sharp increase in imports in 1968 the United States negotiated Voluntary Restraint Agreements with Japanese and European exporters, limiting exports of steel to the United States to a target of 14 million tons in 1969, a target that was allowed to increase gradually in subsequent years. Most studies of this program have found that it raised prices and limited imports through 1972 but was not binding thereafter.[12] The U.S. industry generally believes that the VRAs were mildly successful, although they purportedly led exporters to shift their exports to the higher-valued products to maximize profits under a quantity restraint.

The empirical model in chapter 3 can be employed to provide some evidence of the impact of the VRAs. If the United States were to limit imports

12. For a discussion of the operation of the quotas in different years see Lawrence C. Rosenberg, "Impact of the Smithsonian and February 1973 Devaluations on Imports: A Case Study of Steel," in Peter Clark, Dennis Loque, and R. J. Sweeney, eds., *The Effects of Exchange Rate Adjustments,* Proceedings of a Conference Sponsored by the Office of the Assistant Secretary for International Affairs, U.S. Department of the Treasury (GPO, 1977), pp. 432–34; and James M. Jondrow, "Effects of Trade Restrictions on Imports of Steel," in *The Impact of International Trade and Investment on Employment* (GPO, 1978), pp. 11–25.

Table 5-2. *Residuals from Import Price Equation 3-2 during Voluntary Restraint Agreements*
Dollars per net ton

Year	*Hot-rolled sheet*	*Cold-rolled sheet*	*Plate*	*Bars*	*Structurals*
1968	−3.94	3.61	−1.43	−5.34	−5.48
1969	−4.55	−0.58	−3.57	6.11	−3.42
1970	−1.59	−4.06	−0.81	−10.47	−7.17
1971	8.85	10.70	6.24	9.89	9.56
1972	7.80	7.50	2.43	6.76	8.06
1973	−12.81	−16.70	−13.50	−28.97	−21.64

Source: Author's calculations.

by a global quota and to allow exporters to compete for the limited market, it might be expected that import restrictions would lower import prices if export supply has a positive price elasticity. But the VRAs gave Japan and Europe fixed shares of the quota. Given that import restrictions should have raised U.S. domestic prices, the Japanese and Europeans—freed from the necessity of competing for the U.S. market—might have been expected to reap windfalls from the U.S. policy of protectionism. Indeed, Jondrow's results suggest that this is precisely what happened; export prices rose substantially, as much as 20 percent, owing to the quotas.[13]

The results of the empirical model in chapter 3 fail to confirm Jondrow's findings. While Jondrow's model is estimated for all steel mill products as a single category, my results are for individual product categories, thus avoiding some of the problems caused by the product-mix shifts that may have been induced by the quotas. If the VRAs affected export prices significantly, large positive residuals would have resulted from my cost- and demand-based export price equations (equation 3-1). In fact, this did not occur, as table 3-4 demonstrates. Only in 1973 were residuals generally positive, but in 1973 the quotas were not binding. In 1972 four out of five residuals were actually negative.

The import price equations provide quite a different story. The residuals from the import equations (equation 3-2) clearly rose in 1971–72 from their 1968–70 levels. As the VRAs went into effect in 1969 U.S. import prices were below their predicted values (see table 5-2). By 1971 these residuals had turned strongly positive, and they remained positive until the speculative boom of 1973. Therefore, if it is reasonable to presume that the VRAs caused

13. Jondrow, "Effects of Trade Restrictions on Imports of Steel."

this increase in import prices relative to world export prices, this effect peaked in 1971–72. The minimum estimate of this effect in table 5-3 is taken to be the average of the residuals in 1971–72. The maximum estimate is the difference between this average and the average value of the residuals in 1968–70.

Increases in U.S. import prices feed into U.S. producer prices in the model delineated in chapter 3. As import prices rise, U.S. producers are able to increase their price-cost margins. The estimates of this price-price effect are found in table 3-7. The coefficients of *USIMPRICET* must be multiplied by the assumed increase in import prices to measure the induced effect on U.S. producer prices. The impact varies from a minimum of 46 cents a ton for hot-rolled bars to a maximum of $1.42 for hot-rolled sheet.

The full effect of the VRAs on domestic steel prices may not be registered through increases in import prices. If imports are restrained beyond the level importers would demand at the prevailing import prices, U.S. producers would enjoy the power to raise prices even more than the rise in import prices would suggest. The dummy variable for 1971 assumes a significantly positive coefficient in the U.S. producer price equations (table 3-7) for cold-rolled sheet, structurals, and plate. Moreover, the significantly negative coefficients for bars and hot-rolled sheet in 1969–70 disappear in 1971–72. Therefore, it appears that the 1969–72 period saw increases in U.S. producer prices not otherwise explained by changes in cost, demand, or import prices. This rise in prices can be attributed (at some risk) to the VRAs as well. This imputation appears in table 5-3.

The estimated effect of the VRAs on the import prices of the five major product categories ranges from $4.34 to $11.68 a ton. On the average, the five major categories of steel mill products cost U.S. importers a minimum of 6.3 percent and a maximum of 8.3 percent more because of the VRAs.

The effect on U.S. producers' prices was not so dramatic. If one accepts the estimates of equation 3-3 as valid, a small proportion of the import price increases is transmitted into domestic realized prices, but the additional effects from the restraint of import quantities are more substantial. The average total increase due to the VRAs is estimated to be between 1.2 and 3.5 percent. Obviously this is much less than Takacs's estimate.[14] She finds that steel prices would have been 13 to 15 percent lower in the absence of the VRAs because of her assumption of perfect substitutability of imports and domestic steel and of domestic monopoly in U.S. steel production. Neither assumption

14. Wendy Takacs, "Quantitative Restrictions in International Trade" (Ph.D. dissertation, Johns Hopkins University, 1975).

Table 5-3. *Estimated Effect of Voluntary Restraint Agreements on Domestic and Imported Prices, 1971 and 1972*

Item	*Hot-rolled sheet*	*Cold-rolled sheet*	*Bars*	*Structurals*	*Plate*	*Average*[a]
			Dollars per net ton			
Change in U.S. import prices, excluding import charges						
Maximum	11.68	9.44	6.28	11.56	14.17	10.50
Minimum	8.32	9.10	4.34	8.33	8.81	8.03
Induced change in U.S. producers' prices[b]						
Maximum	1.42	1.08	0.67	0.78	1.26	1.08
Minimum	1.01	1.03	0.46	0.51	0.79	0.84
Total effect on U.S. producers' prices						
Maximum	5.45	4.87	13.57	3.53	3.50	6.00
Minimum	1.01	2.59	0.46	3.31	3.03	2.02
			Percent			
Change in U.S. import prices, excluding import charges						
Maximum	10.1	6.8	5.1	9.3	11.1	8.3
Minimum	0.7	1.5	0.2	1.9	1.7	6.3
Total effect on U.S. producers' prices						
Maximum	3.9	2.8	7.3	2.1	1.9	3.5
Minimum	0.7	1.5	0.2	1.9	1.7	1.2

Source: Author's calculations.

a. Weights are 0.256 for hot-rolled sheet, 0.319 for cold-rolled sheet, 0.154 for bars, 0.121 for structurals, and 0.150 for plate.

b. Assumes that delivered import prices rise by 1.1 times the increase in f.o.b. import prices, reflecting duty, insurance, and interest.

is borne out by the model in chapter 3. Jondrow assumes that U.S. producers' prices are unaffected by imports, an assumption also in conflict with the results of chapter 3. He finds, however, that the VRAs raised import prices by 20 percent, or at least double the estimate of this chapter.

The Effect of VRAs on Market Shares

Given an estimated effect of the voluntary restriction program on import prices of 6.3 to 8.3 percent and on domestic producers' prices of 1.2 to 3.5 percent, import shares must have been reduced. The estimated price elasticities of market shares in chapter 3 average −4. Therefore, import shares should have been reduced by 18.6 to 20.2 percent after a lag of a year or two, ceteris paribus. In fact, import shares averaged 17 percent in 1971–72, then fell to less than 13 percent in 1973–74, a 25 percent reduction. The labor negotiations of 1971 undoubtedly buoyed the demand for imports somewhat, but this effect was temporary.

If one assumes that import shares would have been 18.6 to 20.2 percent higher in 1971–72 than actually recorded, imports would have been between 20 and 22 percent of apparent U.S. supply, somewhat above the 18.1 percent penetration of 1978 in the absence of the VRAs. The U.S. firms would have lost about 3.5 million tons in shipments. Thus this program had the effect of increasing capacity utilization by no more than 3.0 percentage points and of raising U.S. producers' prices by 1.2 to 3.5 percent. Certainly neither effect was sufficiently large to raise the return on equity in the industry to a level sufficient to make new investments profitable. The return remained mired in the 5 percent range after taxes until 1973 and 1974 when the world steel "shortage" drove world prices and U.S. producer returns to very high levels.

The Effect of Trigger Prices on U.S. Import Prices

The 1978 trigger prices were based on Japanese unit costs of production plus freight from Japan. They were first announced in January, but they did not become fully effective before May.[15] How much did this policy raise prices above the levels that might otherwise have been projected, given the deterioration in the value of the dollar in 1978–79?

15. The initial Treasury Department release stated that the trigger prices would not apply to shipments embarking before January 3, 1978. Since the lag between the date of departure from Japan or Europe and the clearing of customs in the United States is generally two months and since many trigger prices were not announced for another one or two months, the system was not fully effective until May 1978.

Table 5-4. *U.S. Import Prices, 1976 and 1979*[a]
Dollars per ton

Item	*Hot-rolled sheet*	*Cold-rolled sheet*	*Plate*	*Bars*	*Structurals*
Average import prices					
1976	194.94	233.30	210.81	222.03	210.92
1979	278.31	336.00	294.54	326.31	320.12
Addendum					
Percent change	42.8	44.0	39.7	46.9	51.8

Source: Table A-8.
a. Excluding importation charges.

Prices for imported steel began falling in 1975, stablized in the second half of 1976, and leveled off in 1977 as world demand failed to recover as rapidly as expected. In 1976 free world production was approximately 13 percent below its long-term trend; by 1979 it had fallen to nearly 18 percent below trend. U.S. prices would have stabilized considerably in 1978 because of the severe decline of the dollar, which fell by 34 percent against the yen between the end of 1976 and the end of 1978.[16] This in itself should have raised Japanese export prices in dollars by 18 to 20 percent.[17] In the same period the dollar declined by 22 percent against the German mark, by 14 percent against the French franc, and by 14 percent against the Dutch guilder.[18] Thus imported steel prices in 1978 would have been higher in U.S. dollars unless the continued sluggishness in demand had led to even deeper discounting by exporters. In 1979, however, the dollar rose once again against the yen, relieving this upward pressure on import prices.

A simple examination of the movement of prices for the five major products demonstrates that there has been a substantial recovery of prices from their low 1976 values. These products, which accounted for 55 percent of 1978 carbon steel imports, had an average increase of 45 percent between 1976 and 1979 (table 5-4).

In chapter 3 the determinants of Japanese export prices and U.S. import prices were estimated for 1956–76. Since there was a break in some of the Japanese export price equations in the early 1960s for most products, it is desirable to reestimate them for 1962 through 1976 in order to make projections for 1978. These estimates appear in appendix B. In order to use the results in appendix B to estimate the effect of trigger prices, one requires estimates

16. International Monetary Fund, *International Financial Statistics,* selected issues, 1976–80.
17. Approximately 55 percent of the cost of Japanese steel reflects domestic value added.
18. IMF, *International Financial Statistics,* selected issues, 1976–80.

Table 5-5. *Changes in Japanese Costs, 1976–79*
Dollars

Input	*(1) Relative use*	*(2) Increase in price*	*(3) Contribution to increase in cost per ton of steel, (1) × (2)*
Iron ore	1.542	3.97 per net ton	6.12
Coal	0.700	8.04 per net ton	5.63
Scrap	0.252	36.31 per net ton	9.15
Electricity	0.472	15.84 per Mkwh	7.48
Oil	0.096	24.52 per net ton	2.35
Total materials			30.73
Labor	. . .	3.35 per hour	. . .

Source: Bank of Japan, *Wholesale Price Index*, selected issues; and Organisation for Economic Co-operation and Development, *Main Economic Indicators* (OECD, April 1980).
Mkwh = thousand kilowatt hours.

of Japanese costs for 1979. The increases in labor and materials costs between 1976 and 1979 appear in table 5-5 (evaluated at an average exchange rate of 219.2 yen = $1). These input price increases add between $46 and $65 a net ton to Japanese costs depending on the product. These cost increases and the changes in the demand variables are inserted into the estimates of equation 3-1 in appendix B to generate predictions of 1979 export prices. In turn, these projections are inserted into the estimates of equation 3-2 (table 3-5) to generate predictions of import prices. The results are displayed in table 5-6.

There are two possible procedures for analyzing the effects of the trigger prices. Predicted import prices may be generated for both 1976 and 1979, and the differences between these values could be compared with the difference in actual prices for the two years. Alternatively, predicted 1979 prices could be compared with actual prices. The first option allows one to assume that departures from predicted prices in 1976 would carry forward to 1979. The second method assumes that prices would have returned to predicted levels by 1979; hence any deviations between actual and predicted prices could be ascribed to trade policy.

The two methods give rather similar results, in large part because 1976 prices for the sheet products and structurals were very close to predicted values. The increase in prices between 1976 and 1979 are substantially greater than would have been predicted in the absence of trigger prices. Bars, structurals, and cold-rolled sheet each rise by $35 to $64 a ton more than the equations would have predicted. The average increases across the five products

Table 5-6. *Effect of Trigger Prices on U.S. Import Prices, 1979*[a]
Dollars per net ton

Item	*Hot-rolled sheet*	*Cold-rolled sheet*	*Plate*	*Bars*	*Structurals*
Predicted price					
1976	201.87	237.18	225.61	196.79	208.36
1979	254.37	301.47	292.36	261.97	276.36
Change in predicted price	52.50	64.29	66.75	65.18	68.00
Actual price					
1976	194.94	233.30	210.81	222.03	210.92
1979	278.31	336.00	294.54	326.31	320.12
Change in actual price	83.37	102.70	83.73	104.28	109.20
Impact of trigger prices					
Change in actual price minus change in predicted price[b]	30.78	38.41	16.98	39.10	41.20
Actual 1979 price minus predicted 1979 price[c]	23.94	34.53	2.18	64.34	43.76
Addendum					
1976–78 weight (share of carbon steel imports)	0.205	0.272	0.198	0.152	0.173

Source: Author's calculations.
a. Import prices are net of importation charges.
b. Weighted average of change in actual price − change in predicted price = \$26.06 (9.1 percent).
c. Weighted average of actual 1979 price − predicted 1979 price = \$32.08 (11.5 percent).

are 9.1 percent, or 11.5 percent more than would have occurred without protection, depending on the method used. A simple average of the two methods yields an estimate of the impact of trigger prices of 10.3 percent by 1979, or about a 5 percent increase a year in import prices. This is substantially less than the claims made by importers during 1978–79, primarily because these latter estimates fail to account for price changes that would have occurred through depreciation of the dollar.

It is possible that the estimates above of the price-enhancing effects of trigger prices might be understated because of the central role ascribed to the Japanese. As the yen appreciated to 180 to the dollar at one point in 1978, Japan probably ceased to be the most efficient exporter to the United States. Canada could probably deliver steel at lower cost to the Great Lakes than the Japanese, and a number of less developed countries may also have had lower costs. But none of these countries was likely to have sufficient capacity to lead a competitive surge against the Japanese in 1978, and by 1979 the yen had fallen to nearly 220 to the dollar. Thus by 1979 Japan undoubtedly was the low-cost producer once again. Therefore, using the Japanese export equation is valid for analyzing 1979 prices.

Table 5-7. *Effect of Trigger Prices on U.S. Producer Prices, 1979*[a]

Item	Hot-rolled sheet	Cold-rolled sheet	Bars	Structurals	Plate
Change in import price, excluding import charges (dollars per net ton)	23.94 to 30.87	34.53 to 38.41	39.10 to 64.34	41.20 to 43.76	2.18 to 16.98
Coefficient from equation 3-3	0.111	0.104	0.0966	0.0618	0.0815
Effect of import price on U.S. producer (dollars per net ton)[b]	2.93 to 3.76	3.95 to 4.39	4.16 to 6.84	2.80 to 2.97	0.20 to 1.52
1976–79 weight (share of domestic carbon steel shipments)	0.275	0.328	0.153	0.102	0.142

Source: Author's calculations.
a. Weighted average change in U.S. producer prices due to trigger price system: $3.05 to $4.04 per net ton (0.8 percent to 1.1 percent).
b. Assumes that delivered import prices rise by 1.1 times the increase in f.o.b. prices.

The Effect of Trigger Prices on U.S. Producer Prices

The 10 percent impact of the trigger prices on 1979 U.S. import prices may be translated into an effect on U.S. producer prices by utilizing equation 3-3 from chapter 3 and the estimates of the effects on U.S. import prices in table 5-6. The total effect, displayed in table 5-7, is between $3.05 and $4.04 a net ton, or 0.8 percent to 1.1 percent of the average price of these carbon steel products in 1979. Predictably the greatest impacts are on cold-rolled sheet and hot-rolled bars, while all other products evidence small increases in price.

In 1979 the producer price index for finished steel products rose by 10.7 percent, compared with 9.9 percent in 1977. Similarly, realized carbon steel prices rose by 10.6 percent in 1978 according to Census Bureau data. But stronger demand and accelerating general inflation should have led to acceleration in steel prices anyway. Increases in raw materials and labor costs for the steel industry accelerated by 0.5 percent in the 1977–78 period, compared with 1976–77. Thus all the increase in the producer price index might well be attributed to cost and demand pressures. In 1979 steel prices rose by another 10.5 percent, substantially less than the rate of increase for all finished or intermediate products in the producer price index. The cost of materials and labor rose by more than 11 percent. Therefore, it seems unlikely that the trigger prices could have increased domestic producer prices by very much.

The evidence that is lacking to reach a more definitive judgment is the magnitude of the reduction in discounting from list prices that might have

Table 5-8. *Changes in Imports and Import Prices, Exclusive of Importation Charges, 1974–79*

Product	*Percent change in imports*				*Percent change in import price*				
	1975–76	*1976–77*	*1977–78*	*1978–79*	*1974–75*	*1975–76*	*1976–77*	*1977–78*	*1978–79*
Hot-rolled sheet	8.5	63.2	−2.0	−17.6	−8.1	−8.8	5.1	13.8	19.3
Cold-rolled sheet	13.7	42.3	−6.6	−25.6	−15.4	−1.8	8.9	12.4	17.7
Hot-rolled bars	−11.9	88.0	−15.4	−25.3	−4.2	−21.0	−1.0	18.9	24.9
Structurals	62.6	27.5	6.0	3.1	−3.1	−21.0	−1.1	21.4	24.0
Plate	15.0	35.6	36.4	−38.5	−3.4	−21.8	1.6	10.4	24.6

Source: Based on data in U.S. Bureau of the Census, *U.S. and General Imports—TSUSA—Commodity by Country of Origin*, FT246 (Government Printing Office, 1975), and subsequent issues through 1979.

occurred in 1979 owing to trigger prices and the increase in product demand caused in part by them. It is possible that the estimate above of a less than 1 percent effect of trigger prices is a reflection of narrowing discounts, but this is simply speculation until realized price data are pieced together from Department of Commerce reports in 1981.

The estimated overall effect on U.S. steel prices is simply the weighted average of the increase in imported and domestic prices attributed to the import restrictions. Assuming that imports accounted for 15 percent of U.S. shipments in 1979 and domestic producers for the remaining 85 percent, the maximum effect may be calculated as 0.15×11.5 percent plus 0.85×1.1 percent, or 2.7 percent. In the absence of trigger prices U.S. steel prices would have averaged at most 2.7 percent less than recorded in 1979, or about \$11.50 a ton below realized market prices.

The Effect of Trigger Prices on Market Shares

Given the lags that operate in translating relative-price changes into shifts in market shares, it would be very difficult to measure the impact of the trigger price system on the share of imports in the U.S. market. Imported prices of all the mill products analyzed in this study fell substantially in 1975 and 1976. The effects of these declines were registered in market shares in 1976 and 1977. For sheet products the largest declines in imported prices occurred during 1975 (table 5-8), with the surge in imports following by two years in 1977. The import prices of bars and structural steel products fell sharply in 1976 after a modest decline in 1975 and imports followed after a shorter lag, rising by approximately 80 percent between 1975 and 1977.

Plate imports continued to rise through 1978 despite a reversal in import price in 1977, perhaps as a result of anticipated increases in trigger prices.

The changes in import quantities or shares in 1978 could not be ascribed to trigger prices. In the first place, trigger prices did not begin to have an effect on imports until midyear. Before that, anticipations of rising prices due to the policy announced in January would have caused substantial anticipatory buying. Moreover, import prices accelerated very rapidly after April. In the second place, except for structurals one would not expect rising import prices to affect the share of the market appropriated by imports until 1979, given the estimates of the market share equation (3-4) in chapter 3.

Given the firming of import prices in 1977 and the sharp increases in import prices in 1978 and 1979, one would have expected imports to slow in 1978 and to drop precipitously in 1979. As table 5-8 demonstrates, this is precisely what happened for everything but structurals. Even structural imports failed to grow very much in 1978 and 1979.

The overall impact of trigger prices on market shares can be deduced from the information on relative price effects and the coefficients of equation 3-4. The relative price of imports was increased by trigger prices by a maximum of 10.3 percent (1.115/1.011) in 1979. Given an estimated price elasticity of market share averaging −4, this suggests that the share of the U.S. carbon steel market accounted for by imports was 41 percent lower than it would have been (once the full effects are realized). If, for instance, imports would have remained at a 0.18 share of the U.S. market, trigger prices should have reduced this share to 0.11.

Imports did fall in 1979 in response to higher relative import prices in 1978. Import prices rose by 14.7 percent in 1978 and by another 22 percent in 1979. Thus less than 30 percent of the rise in import prices in these two years should be ascribed to trigger prices. Domestic prices of carbon steel products rose by 21.8 percent in 1978 and 1979. The increase in relative import prices in 1978 and 1979 was 15 percent, which—other things being equal—should eventually translate into a 60 percent reduction in the market share going to imports. Given a 15 percent time trend in the import share, the overall result should be approximately a 28 percent reduction in market share from the 1977 level of 0.18 to approximately 0.13. In fact, imports declined to a 0.15 share in 1979 but began to rise once more in 1980. In short, while trigger prices had an effect on prices and output in the U.S. steel market, the decline in the value of the dollar and general inflation in basic world materials costs had a much more powerful influence.

Summary

Whatever the merits of developing a trade policy for increasing U.S. self-sufficiency in steel, there have been two bouts of trade protection since 1968. Each has raised import prices far more than domestic prices. The VRAs raised import prices by 6.3 to 8.3 percent, while domestic prices were increased by only 1.2 to 3.5 percent. Similarly the 1978–79 trigger prices raised import prices by an estimated 9.1 to 11.5 percent while allowing domestic producers to raise prices by a modest 0.8 to 1.1 percent. This latter estimate may appear surprisingly low, but corroboration is to be found in the fact that the reported increase in the prices of steel mill products in 1978 and 1979 was only average for the economy during that period. Given the decline in the value of the dollar, the weak market in 1977, and the escalation in the cost of materials and labor for the steel industry in 1978 and 1979, it is difficult to see how the trigger prices could have had much influence on U.S. producers' prices. The major beneficial effect of trigger prices for U.S. steel companies lay in their suppression of import demand. But did the protection move the industry closer to being able to modernize through the construction of new plants? This question is addressed in the next chapter.

VI

Alternative Government Policies for Maintaining U.S. Self-Sufficiency

In the current economic climate the expansion of steel capacity in the United States seems quite unlikely. Since the recession of 1958, the date at which declining U.S. comparative advantage began to be registered, there has been virtually no capacity growth in the U.S. industry. In 1960 the United States had 143 million tons of raw steel capacity; today it has 153 million tons of "capability."[1] The annual rate of growth of capacity over these twenty years has been less than 0.5 percent. The remainder of the growth in steel consumption has been satisfied by imports. It is possible to consider a range of policies that might restore the U.S. producers' position in their home market. Such policies may be unrealistic, but an analysis of them provides a benchmark against which to compare the current situation.

How Much Capacity?

Throughout most of the late 1950s the U.S. steel industry operated at considerably less than full capacity. Obviously more aggressive price competition might have gained the industry a larger share of its own home market, reducing imports somewhat. But as the late 1960s arrived, the industry's ability to supply its own market began to recede. Capacity expanded very little as U.S. consumption expanded steadily. By 1968–69 the industry's capacity was only 152 million net raw tons while U.S. consumption averaged 153 million net raw tons. The industry had clearly lost its ability to supply its own market at the going prices.

For the industry to have been able to continue to supply the U.S. market through the 1973–74 "shortage" period, it would have had to build two additional large integrated plants in the 1960s, each capable of producing a

1. See table 2-4. The estimate for 1960 capacity is based on interpolations between peak supply periods. The 1980 estimate is the American Iron and Steel Institute's raw-steel "capability" measure.

Table 6-1. *U.S. Production, Consumption, and Trade in Steel Products*
Millions of net tons of raw steel

Year	*Capacity*	*Production*	*Imports*[a]	*Exports*[a]	*Apparent supply*	*Capacity with two new plants of 6 million tons each*[b]
1965	148.2	131.5	13.9	3.3	142.1	154.2
1966	149.4	134.1	14.4	2.3	146.1	155.4
1967	150.6	127.2	15.3	2.3	140.2	156.6
1968	152.2	131.5	24.0	2.9	152.6	158.2
1969	152.8	141.3	18.7	6.9	153.1	164.8
1970	153.8	131.5	17.9	9.4	140.0	165.8
1971	154.8	120.4	24.4	3.8	141.0	166.8
1972	156.2	133.1	23.6	3.9	152.8	168.2
1973	156.7	150.8	20.1	5.4	165.5	168.7
1974	157.4	145.7	21.3	7.8	159.2	169.4
1975	157.4	116.6	16.0	4.0	128.6	169.4
1976	157.7	127.3	19.1	3.6	142.8	169.7
1977	158.1	125.3	25.7	2.7	148.3	170.1
1978	156.0	137.0	28.2	3.2	162.0	168.0

Source: Annual statistical reports of the American Iron and Steel Institute and table 2-4.
a. 0.75 yield equivalent.
b. One beginning in 1965 and one in 1969.

sustained 6 million net raw tons a year from 6.4 million tons of nominal capacity (95 percent capacity production). As table 6-1 demonstrates, this would have provided domestic capacity of nearly 169 million net raw tons in 1973, compared with U.S. consumption of approximately 165 million net raw tons. It would have increased 1978 capacity to 168 million tons, assuming no further plant closures. How might such an expansion be induced by government policy?

Stimulating New-Plant Investment

As chapter 4 demonstrated, the economics of new-plant construction in the United States are extremely unfavorable. If, as appears increasingly likely, brownfield (round-out) expansions cost $700 to $800 per net raw ton, even these forms of expansion will not occur unless cost-price relationships are altered substantially. In 1978 dollars the cash flow from a new plant would be approximately $96 per net ton of finished products (an average mix of

Table 6-2. *Parameters for New-Plant Analysis*

Year	*Before-tax cash flow (dollars per finished ton)*	*Cost of plant (dollars per finished ton)*	*Moody's Aa bond rate, r_{Aa} (percent)*	*Treasury bill rate—one-year maturity, r_{tb} (percent)*	*Estimated cost of equity capital, before taxes, ρ_e (percent)*	*Gross domestic product deflator (1972 = 100)*	*Estimated cost of capital, ρ (percent)*
1961	...	...	4.4	2.4	...	69.2	...
1962	...	...	4.3	2.8	...	70.5	...
1963	...	...	4.3	3.2	...	71.6	...
1964	...	...	4.4	3.5	...	72.7	...
1965	51.00	376.25	4.5	4.0	15.6	74.3	12.8
1966	52.00	387.50	5.1	4.9	16.5	76.8	13.5
1967	51.00	400.00	5.5	4.3	17.0	79.0	13.9
1968	51.00	417.50	6.2	5.3	17.7	82.6	14.6
1969	49.00	437.50	7.0	6.7	18.8	86.8	15.6
1970	51.00	463.75	8.0	6.5	19.5	91.4	16.3
1971	55.00	493.75	7.4	4.3	19.5	96.0	16.4
1972	54.00	513.75	7.2	4.1	19.0	100.0	16.1
1973	54.00	535.00	7.4	7.0	19.1	105.7	16.2
1974	76.00	601.25	8.6	7.9	19.7	115.6	16.7
1975	72.00	697.50	8.8	5.8	20.3	126.8	17.2
1976	67.00	753.75	8.4	5.0	20.7	133.3	17.6
1977	77.00	885.00	8.0	5.3	20.0	141.1	17.1
1978	96.00	937.50	8.7	7.2	19.7	151.5	16.9

Source: Cash flows based on unit material costs and labor costs assuming 1.5 percent productivity growth. See appendix A for input price data. See chapter 4 for 1978 estimate of cash flow. Cost of plant based on data in table 4-3. All other data are from *Economic Report of the President, January 1979.*

$$\rho = \frac{R}{4} + 3/4(\rho_e) = \text{weighted average cost of capital,}$$

$$R = \left[\frac{r_{Aa} + r_{Aa,-1} + r_{Aa,-2} + r_{Aa,-3}}{4}\right] \times 100 = \text{cost of debt capital,}$$

$$\rho_e = \left\{\left[r_{tb} + r_{tb,-1} + r_{tb,-2} + r_{tb,-3}\right] 1 + 0.06\right\} \div 0.6 = \text{cost of equity capital}$$

carbon steel excluding tubular and wire products). Even if this were assumed to grow at 6 percent a year over the life of the plant owing to inflation, the present value of a new plant would be no more than $823. This is 12 percent below the cost of building the mill. The average price of steel would have to rise by $13 a net ton while operating costs remained constant to make such an investment attractive.

The degree of subsidy required to induce expansion obviously depends on the expectations of steel executives. If they expect price-cost margins to rise over the life of the plant at the rate of recent inflation, they would require less subsidy to build capacity at current construction costs than if they believed that the rate of increase would more closely approximate recent trends in the cash flow per ton achieved by the industry (table 6-2). In the analysis that follows, the reduction in the cost of building a new flat-rolled carbon steel plant, whether by direct subsidy or tax reform, is calculated under two assumptions about the expected growth in before-tax cash flow per ton from a new plant. These assumptions are that cash flow will grow at a rate equal to (1) the average growth rate in the gross domestic product (GDP) implicit deflator over the past four years and (2) zero. The relationship between the ex ante present value of the new plant per ton of finished project capacity under each assumption and the cost of that capacity is presented in table 6-3.

It is clear that even the more optimistic scenario, that cash flow will increase for twenty-five years at a perpetual rate equal to the recent rate of increase in the GDP deflator, does not generate sufficient future flows to amortize the 1978 cost of a new plant. The more conservative, and probably more realistic, assumption is that steel executives expect a rate of increase somewhere between zero percent and the recent inflation rate. In fact, there is a strong view expressed within the steel companies that planning must be based on the zero cash flow growth assumption because government policy will probably limit all price increases to the level of cost increases for an indefinite period.

Table 6-2 contains the assumptions used in the analysis of the 1966–78 hypothetical investment decisions. The cost of capital is assumed to be equal to a weighted average of debt and equity costs with weights of 0.25 and 0.75, respectively. The cost of equity capital is 6 percentage points more than the four-year average of three-month commercial bill rates. The cost of debt is the four-year average of the Moody's Aa rate. Plant life is assumed to be twenty-five years. The tax rate on equity is assumed to be 0.40 throughout the period.

The results of the discounted cash flow analysis are reproduced in table

Table 6-3. *The Value of Investment in a New Flat-Rolled Carbon Steel Plant at 90 Percent Capacity, 1966–78*
Dollars per net ton of finished-product capacity

		Present value of facility with annual cash flow growth equal to:	
Year	*Cost of facility*	*Average increase in GDP deflator*	*Zero percent*
1966	387.50	430.00	372.00
1967	400.00	421.00	355.00
1968	417.50	421.00	340.00
1969	437.50	396.00	308.00
1970	463.75	407.00	308.00
1971	493.75	451.00	330.00
1972	513.75	450.00	329.00
1973	535.00	450.00	327.00
1974	601.25	656.00	448.00
1975	697.50	651.00	412.00
1976	753.75	596.00	376.00
1977	885.00	718.00	444.00
1978	937.50	874.00	560.00

Source: Author's calculations.

6-3. Each of the two assumptions generates a negative net present value for the 1978 investment, but the 6.8 percent projected growth rate in cash flow that is obtained from projecting the last four years' growth in the GDP deflator is obviously very optimistic. In 1978 the discounted present value of a new investment was between 6.8 and 40.3 percent below the cost of the facility. If the plant operated at an average of 95 percent of capacity for its entire twenty-five-year life, an optimistic assumption, the present value would fall to 11.4 to 43.3 percent below the plant's cost. Even these economics are optimistic because 1978 was a year of an unusually low value of the dollar in terms of Japanese yen. In 1977, for example, a plant was worth only 48 to 77 percent of its cost.

For the analysis that follows, it will be assumed that investors expected a cash flow growth rate between the two extremes in table 6-3. The present value is assumed to be 27 percent below the 1978 cost of the plant, the midpoint of the 95 percent capacity results referred to above, ignoring the artificial stimulus provided by trigger prices. The cost of the plant would have to be reduced by 27 percent, the cost of capital reduced by more than 3 percentage points, or the price of steel would have to rise by more than 9 percent immediately (in 1978) and continue to increase at the inflation rate thereafter if a new plant were to be profitable under these assumptions. This may be a more optimistic conclusion than that reached by the Steel Tripartite Committee in 1980.[2] The committee foresees the need to raise prices by nearly 4 percent and to reduce corporate tax liabilities just for the industry to modernize its existing facilities.

Trade Protection

Clearly trade protection could be employed to limit imports, raise prices, and thereby create a favorable climate for domestic investment. To raise domestic steel prices sufficiently to encourage new investment, trade protection would have to be rather severe. Given that the trigger price policy raised domestic prices by approximately 1 percent in 1978, the protection required for more than another 8 percent increase would be substantial. Just how this could be accomplished under U.S. law is not important for this analysis.[3] Presumably quantitative controls on imports would be required. If two new plants were to be built, each producing approximately 6 million net tons of raw steel, imports would have to be reduced to virtually zero, given the average import level over the past decade. The cost to U.S. consumers and the welfare cost to the United States generally would depend on the method of achieving this reduction.

Assume that trade protection had been advertised several years before 1978 so that two new plants with annual sustainable output of 6 million finished tons of carbon steel would have opened in 1978. These plants would have produced 9.6 million net tons of finished carbon steel products at an 80 percent yield. In 1979 domestic firms' shipments of carbon steel were 90 million net tons from approximately 99 million tons of capacity. The new plants would have increased capacity to 108.6 million tons of carbon steel products. Given total imports of 17 million tons, it is obvious that imports would have had to be eliminated altogether to allow as many as two additional plants to be built and operated at capacity without retirements.[4] Given 1979 carbon steel prices of $423 a net ton, the increase in domestic steel prices required to facilitate this construction in 1979 would be $34 a net ton. At a minimum this policy would transfer $3.5 billion annually from consumers to producers.

2. See "Report to the President by the Steel Tripartite Committee on the United States Steel Industry" (September 25, 1980).

3. See chapter 7 for a discussion of this point.

4. In chapter 7 a more realistic scenario is analyzed more intensively.

The welfare cost of protection of this variety is more difficult to calculate because one needs to estimate the additional investment that would be induced in the form of new plants and renovations of older ones. These latter investments would look much more attractive with prices perpetually 9 percent higher than in the absence of trade protection.

The welfare loss from the new plants is relatively easy to calculate. Each plant with 6 million tons of raw steel capacity and a steady stream of 4.8 million tons of output would cost 6 million times $750, or $4.5 billion. The present value of these investments without trade protection would be $3.3 billion each. Hence the welfare loss of subsidizing the construction of two plants of this variety would be $2.4 billion. Some of this excess might be a payment for the social externality for national defense suggested by Thompson,[5] but this would account for less than 30 percent of the welfare cost.

The welfare cost of additional investments made at existing sites cannot be calculated with similar precision. In recent years capital expenditures in the iron and steel industry have averaged somewhat more than $3 billion (1978 dollars).[6] A larger pretax profit margin caused by the 8 percent increase in price would undoubtedly trigger some additional renovation projects. In 1977 the major companies had a backlog of several billion dollars in potential projects of this sort, which were not funded because of capital constraints and their failure to generate a sufficient rate of return. Simply raising output prices would not improve the economics of these projects unless they offered increments to output. Just what percentage of the capital invested in such projects might generate expanded output is unknown, but it would probably be small. To the extent that higher prices unlocked a cash flow constraint, the additional investment generated would not create economic waste. Hence, it is unlikely that the welfare losses from renovations spurred on by higher output prices would be very large. The major welfare cost of higher steel prices, within the steel industry, would derive from the uneconomic construction of new plants.

Reducing the Cost of Capital

To build new steel capacity in the United States at present requires an annual return to capital approximately 30 percent above current levels. That

5. Earl A. Thompson, ''An Economic Basis for the 'National Defense Argument' for Aiding Certain Industries,'' *Journal of Political Economy*, vol. 87 (February 1979), pp. 1–36.

6. Information supplied by the U.S. Bureau of Economic Analysis.

is, an investment made in 1979 would require annual cash flows per ton of $34 above those obtainable in 1979. A capital subsidy equal to 30 percent of outlays or an equivalent reduction in the cost of the capital through a change in the tax laws would therefore be the minimum required to induce new investment in the steel industry. Accelerated depreciation, a sharply increased investment tax credit, or lower corporate income taxes for this industry alone (an unlikely possibility) could be utilized to provide this reduction in the cost of capital from approximately 17 percent to 12.5 percent. The welfare cost of this form of tax relief would be similar to that provided by import protection, but the transfer of wealth would be much smaller, because the prices of steel products would not be increased.

The industry has advocated the passage of the Capital Cost Recovery Act, which would shorten the useful life of plant and equipment for income tax accounting to ten and five years, respectively. If the after-tax cost of capital were unaffected by the application of this change in the tax laws to all industries, this provision would increase the present value of a steel plant by only 11 percent, far less than required to make a plant viable at current prices and input costs.

The cost to society from a capital subsidy or tax reduction would be related solely to the investment in otherwise uneconomic facilities. The cost would include the $2.4 billion in 1978 present value from building the two new 6-million-ton plants, but it would also include the additional investment in the rehabilitation of existing steel mills, the magnitude of which is not calculable at present.

Certainly a 30 percent reduction in the cost of capital could not and should not be limited to new plants. Some investment in older plants may be more economic than greenfield outlays; hence if the latter are to be subsidized, the former should be encouraged as well. It would not be surprising if a 30 percent reduction in capital costs would increase the spending on existing mills by $1 billion to $2 billion a year. Most of this spending would not add measurably to capacity; rather, it would reduce the cost of production for existing capacity. Somewhere between zero and 30 percent of each investment would reflect an expenditure in excess of current market value, or zero to $600 million a year. In short, the cost to society of increasing investment in this fashion is similar to the cost of trade protection, but steel *prices* are not affected in any important way. The transfer of wealth from consumers to producers equal to $3.5 billion a year under trade protection would be avoided. Moreover, if the private cost of capital is now too high because of current tax policy, some of the additional expenditures would not reflect a social welfare loss.

Inventory Accumulation

If the purpose of increasing U.S. self-sufficiency in steel is to be able to gird better for a major national defense effort, the additional steel supply required could be provided from stockpiles subsidized by the government. For instance, rather than subsidizing the construction of two new plants producing 12 million tons a year, the government could finance equivalent extra inventories. This would get more costly and burdensome if a much longer emergency were hedged in this fashion. An additional supply of 12 million tons could be held in the form of semifinished billets, blooms, and slabs, which could be rolled as needed in the rolling mills of the integrated and nonintegrated steel companies. In general the capacity in rolling mills exceeds basic steelmaking capacity, and this 12 million tons would only add 7.5 percent to current maximum rolling, but it would more than double current in-process inventories. Finished products could not be held in inventory as usefully because of the diversity of sizes and shapes of these mill products and because they would degrade. Semifinished products might oxidize more than normal if held longer in inventory, but this could easily be rectified in scarfing processes normally performed as these products are rolled into finished products.

A ton of semifinished product represents approximately one-half of the capital costs, less than 40 percent of labor costs, and nearly all of the materials costs (other than energy) required for a finished product. At 1979 prices this semifinished product would be worth approximately $220 a net ton. A 12-million-net-ton inventory would thus be worth $2.6 billion. At a 10 percent discount rate and assuming storage and handling costs of 5 percent a year, the cost of holding this inventory might be about $390 million a year. Obviously, to hold four years' extra supply of 12 million tons annually would require four times this figure, or $1.56 billion annually. This inventory cost is similar to the real resource costs of capital subsidies and trade protection, but the transfer of wealth is less than under trade protection.

A drawback to most government stockpile schemes is that they discourage the private holding of inventories. If the government is to hold an inventory for disposal during an unexpected shortfall of production, private investors are less likely to hold as large inventories since the government is embarked on a course designed to reduce the profits from holding these inventories.[7]

7. For a discussion of the problems with stockpiles see U.S. National Commission on Supplies and Shortages, *Government and the Nation's Resources: Report of the National Commission on Supplies and Shortages* (The Commission, 1977), chap. 7.

This is likely, however, to be less of a problem for materials held only for release during major national emergencies than for those held to stabilize prices. While it cannot be doubted that a government steel inventory would reduce private inventories by some small amount, if it is understood that these stockpiles are not to be released to offset strikes, shipping disruptions, or price inflation, this effect should be minor.

Long-Term Contracts

Many industry observers have opined that buyers would have been better off buying domestic steel in the 1960s and early 1970s so as to induce U.S. steel companies to expand capacity, which would have eased or alleviated the "shortage" of 1973–74. The prescription for U.S. steel buyers is obviously one of long-term contracting: in return for patronage in slack years U.S. producers would supply steel at less than world prices during periods of boom or "shortage."

It is difficult to specify precisely how such an implicit long-term contract might operate. Assume that U.S. firms had expanded capacity in the mid-1960s when such expansion would have been profitable if the additional sales could have been obtained at current transactions prices. According to table 6-3, this possibility disappeared in 1968 or 1969 when the cost of new facilities began to exceed their expected present value at the going price level. But before 1969 such a contract might have been feasible. Would it have been profitable to buyers? To determine the answer it is necessary to compare import prices and a comparable mix of domestic prices over a period from the 1960s through the 1970s.

Imports began to rise sharply in the mid-1960s as import prices, including freight and duty, fell below U.S. producer prices. Beginning in 1964 the weighted average of import prices for the five categories of products examined in this book fell substantially below U.S. producer prices. They remained below U.S. prices until 1973, as table 6-4 demonstrates. Assume that U.S. buyers had negotiated to receive additional tons of steel from 1964 through 1976 at U.S. producer prices precisely equal to subsequent transactions prices in those years. U.S. firms could have profitably expanded capacity had they been assured of this demand at the prices obtained in the 1964–76 period and if cash flows had continued to grow at 3 percent thereafter. Buyers would have saved as much as an average of $61.32 a ton in 1974 but would have paid premiums ranging up to $32.45 in 1976 and $21.79 in 1968. The present value of the savings per net annual ton bought under this long-term contract

Table 6-4. *Relative Prices of U.S. Imports of Five Major Carbon Steel Products, 1956–1976*[a]

Dollars per net ton using current import weights

Year	*U.S. import price (including importation charge)*	*U.S. producer price*	*Difference*
1956	172.20	111.94	60.26
1957	166.86	122.38	44.48
1958	133.21	128.18	5.03
1959	131.13	130.18	0.95
1960	149.93	130.97	18.96
1961	130.73	130.53	0.20
1962	129.94	130.44	−0.50
1963	124.85	130.54	−5.69
1964	119.81	133.16	−13.35
1965	122.24	133.26	−11.02
1966	115.74	133.52	−17.78
1967	115.72	133.88	−18.16
1968	115.49	137.27	−21.78
1969	122.60	141.69	−19.09
1970	145.79	151.56	−5.77
1971	147.24	161.61	−14.37
1972	161.17	173.00	−11.83
1973	192.48	183.10	9.38
1974	302.43	241.11	61.32
1975	296.94	267.48	29.46
1976	248.88	281.34	−32.45

Source: Based on data in appendix A.

a. The five major carbon steel products include cold-rolled sheet, flat-rolled sheet, plate, bars, and structurals.

would have been −$64.91 at a 5 percent discount rate, −$60.38 at a 10 percent discount rate, and −$54.31 at a 15 percent discount rate. Thus buyers would have lost substantially in such an arrangement. For a 12 million net ton "commitment" of this sort, the present value of losses in 1964 would have been between $652 million and $779 million. Obviously the losses would have been even greater if U.S. buyers had been able to shop freely for steel during the 1969–72 VRAs period.

The result above should not be surprising. If the analysis were based on a later period, it would be even less favorable since buyers would have to pay premiums over U.S. producer prices to induce expansion of capacity as a hedge against "shortages." Indeed, at present these premiums might have to be as much as 8 percent of current prices. Since Japanese and other emerging steel exporters in the 1960s were able to build and operate steel mills at lower costs than their U.S. rivals could, competitive prices in world markets, including premium prices during booms and bargain prices during slumps, should have been more attractive than U.S. prices for the incremental output. My analysis shows that this would have been true even when importation charges are added to the exporters' prices.

Summary

Construction of new plants in the United States to maintain U.S. self-sufficiency in carbon steel would require the subsidization of only two new integrated plants of 6 to 7 million tons each. While such plants might have been built profitably before 1969, assuming that buyers would shift their patronage from imports to the U.S. mills at current U.S. producer prices, major reductions in the cost of capital or direct subsidies would now be required to generate new-plant construction. These reductions would have to be enormous, totaling nearly 30 percent of the new-facility cost. For two plants there would have to be a reduction of cost through subsidies or liberalized tax laws (or both) of $2.4 billion.

If it were decided that national security or other requisites of social policy demanded that the United States be virtually self-sufficient in steel, a number of alternatives to this self-sufficiency would be possible. Among these are trade protection, capital subsidies, inventory subsidies, and long-term contracts. Each of the first three could be undertaken by government; the choice among them would hinge on the welfare costs of each and the desired income redistribution. The last policy is a strategy buyers could underwrite, although government could subsidize them to the extent necessary to effect the shift from imports to domestic products.

Each policy designed to increase U.S. productive capacity would involve a loss in producers' surplus equal to the excess of the cost of the facilities over their present value without the subsidy. The passage of the Capital Cost Recovery Act, shortening the depreciable life of assets would not be sufficient. Further capital subsidies would probably encourage modernization of plants in excess of the social value of the prospective cost savings. For carbon steel products alone, trade protection sufficient to make U.S. new-plant investment attractive would transfer $3.5 billion a year from consumers to producers, far more than the cost of the VRAs or the trigger prices.

Subsidized inventory accumulation of semifinished products would represent the best strategy for hedging against a short national security emergency of one or two years. The welfare losses involved in subsidizing capital facilities and raising domestic prices would be avoided by such a strategy.

Finally, long-term contracting for U.S. supplies would occur only if the government subsidized such contracts, since the present value of hedging against a ''shortage'' price of steel in the world market, based on the evidence since 1964, is decidedly negative.

The results of this chapter point to tax reductions, capital subsidies or inventory-subsidies as the best mechanisms for increasing U.S. self-sufficiency in steel during time of crisis. The critical question is whether such self-sufficiency is required in a world of increasing diversity of steel supply. The discussion in chapter 5 suggests that such a policy, whether based on national security or concepts of ''fair'' trade, has little merit. This chapter simply confirms that the policy would be expensive.

VII

The Gainers and Losers from U.S. Trade Protection

The analysis presented in chapter 5 suggests that the two major bouts of U.S. trade protection for the steel industry have had rather small effects on U.S. producers' prices but much larger effects on import prices. Moreover, while the lagged effect on market shares is not inconsequential, trade policy cannot be used to maintain employment at aging, inefficient plants. Thus the employment effects of trade protection in the long run are likely to redound to labor markets in which steel continues to be efficiently produced—largely Great Lakes markets and a variety of small communities throughout the country in which mini-mills are located.

It is finally possible to sum up the distributional and welfare effect of trade protection for the steel industry. The effects of trade restrictions are estimated to be equivalent to that under the 1978–79 trigger price system and of restrictions sufficient to induce new-plant construction in the United States. The results of chapters 3, 5, and 6 are employed in this exercise, borrowing from other studies for a few remaining empirical magnitudes.

The basic model used in estimating the effects of trade policy is detailed in figure 7-1. Trade policy raises the import price of steel by ΔP_m or (for an equivalent quantity restraint) lowers imports by ΔQ_m. The increase in import price shifts the domestic demand curve from D_d to D_d'. The increase in import prices leads domestic suppliers to raise prices by ΔP_d, which in turn raises the import demand schedule to D_f'. The post-protection equilibrium is given by P_m^*, Q_m^*, and P_d^*, Q_d^*.

For the purpose of this analysis "domestic" production and shipments include U.S. exports, which are small and go mostly to other western hemisphere countries. To attempt to include the export market separately, given the U.S. industry's small participation in it, would add needless algebraic and geometric complexity. The model depicted in figure 7-1 assumes competition exists in the domestic market and that the domestic price response to trade protection depends on the elasticity of supply. One can proceed to analyze the effects of trade protection on welfare by following this assumption

Figure 7-1. *The Effects of Trade Protection*

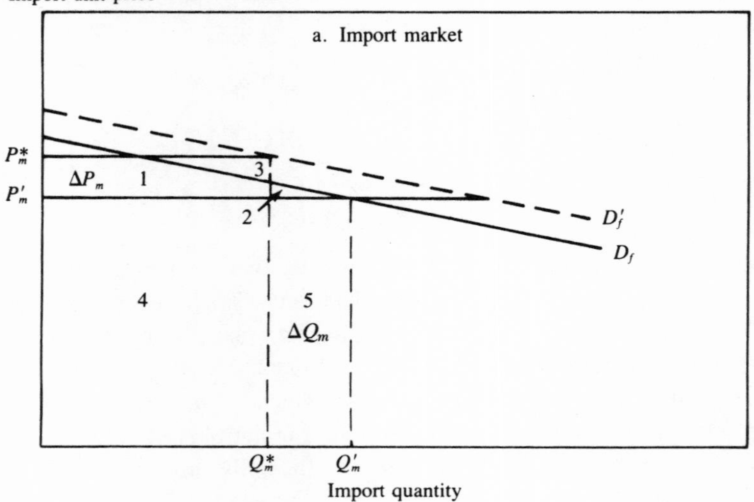

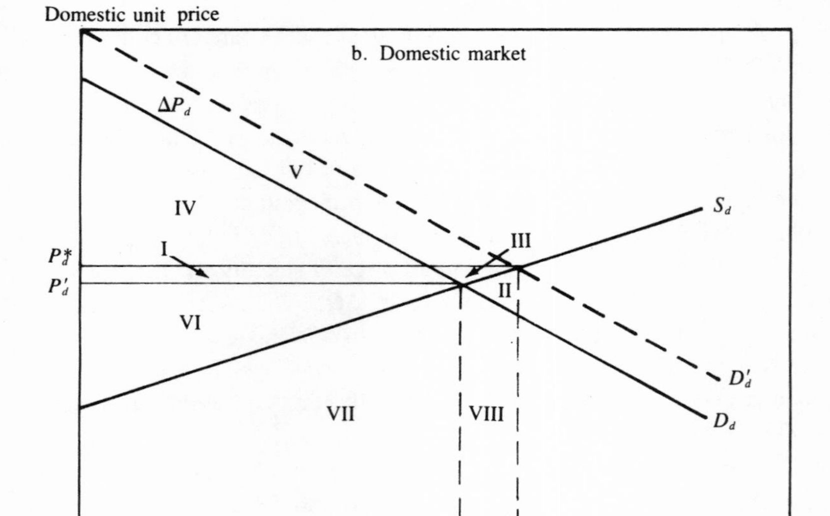

Table 7-1. *Econometric Estimates of Import and Domestic Steel Demand Elasticities*

	Domestic market		*Import market*	
Product	*Own-price elasticity*	*Import price elasticity*	*Own-price elasticity*	*Domestic price elasticity*
Hot-rolled sheet	−0.54[a]	0.31[a]	−2.54 to −10.45	0 to 14.41
Cold-rolled sheet	−1.49[b]	0.77[b]	−5.17 to −7.35	7.29 to 9.78
Bars	−1.30[b]	0.63[b]	−1.09 to −1.30	0
Structurals	−1.99	1.08	−0.54 to −3.26	4.11
Plate	−1.81[b]	1.09[b]	−1.09 to −2.82	3.50

Source: Appendix B.
a. Not statistically significant.
b. Two-stage least squares estimate.

or by assuming that domestic firms are price setters, responding to foreign prices per equation 3-3.

To estimate the effects of trade policies on different groups in the domestic and world economy, it is necessary to have separate estimates of the demand functions for domestic and imported steel and either a domestic supply equation or a domestic price adjustment equation. Estimates of individual product demand curves by ordinary and two-stage least squares appear in appendix tables B-4 and B-5. These results are summarized in table 7-1. The domestic demand equations yield own-price elasticities ranging from −0.5 to −2.0, while the cross-price elasticities (with respect to import prices) are from 0.3 to 1.1. The import equations evidence a much wider range of estimates, with own-price elasticities varying from −0.5 (for structurals) to as high as −10 (for hot-rolled sheet). The cross-price elasticities for imports are also much higher than in the domestic demand curve estimates, ranging from 3.5 to 14.4, where statistically significant.

For the purposes of the simulations that follow, I use the following domestic and import demand equations for the entire carbon steel market:

$$\log Q_d = a_o - 1.5 \log P_d + 0.6 \log P_m, \tag{7-1}$$

$$\log Q_M = b_o + 4.0 \log P_d - 4.5 \log P_m, \tag{7-2}$$

where Q_d and Q_M are domestic shipments (including the small amount of exports) and imports of carbon steel, respectively, and P_d and P_m are domestic producer prices and import prices.

The supply elasticity estimate of 1.38, obtained by Jondrow and Katz, is inconsistent with U.S. firms' pricing behavior.[1] In fact, the behavior of these firms appears more consistent with a short-term supply elasticity of approximately 3.5. This suggests that when demand falls as import prices fall, U.S. producers reduce their prices by less than the price reduction one would expect even in the long run from a competitive industry. Therefore, the industry is clearly not behaving in a manner consistent with competitive theory. The "supply elasticity" of 3.5 provides a representation of the industry's price response to demand shifts but not a measure of the shifting of competitive equilibrium as import market conditions change.

The comparative statics of a change in import prices under different assumptions concerning domestic industry supply conditions can be depicted by redrawing figure 7-1, as in figure 7-2. The long-run competitive supply curve for the industry is shown as the dotted line S_p, while the "price response" curve is a much more price-elastic R_p. If the industry is at P_d', Q_d' before trade protection and is propelled to $[P_d^*, Q_d^*]$ by protection, it remains off its competitive supply curve. The departure depends on the oligopoly pricing pattern of the industry and the ability of workers to obtain supracompetitive wages. If wages were closer to the all-manufacturing average, the opportunity cost of supplying steel would be lower at each rate of output, and the domestic steel prices and the import share would be lower if competitive conditions obtained in the product and labor markets. These considerations do not affect the deductions concerning the effects of trade policy given R_p except for one parameter—the share of factor payments for the additional steel produced that represents rents to the claimants.

Since alloy and specialty steels are excluded from the analysis, only the quantities and prices for imports and domestic producers' shipments of carbon steel in 1979 are required.[2] Domestic shipments in 1979 were 90 million net tons while imports accounted for 17 million tons. The import "share" was therefore approximately 16 percent. The average domestic price is assumed

1. James Jondrow, "Effects of Trade Restrictions on Imports of Steels," in U.S. Department of Labor, Bureau of International Labor Affairs, *The Impact of International Trade and Investment on Employment* (Government Printing Office, 1978), pp. 24–25; and Arnold Katz, "Estimates of Labor Displacement: A Structural Approach," in James Jondrow and others, "Removing Restrictions on Imports of Steel" (Arlington, Va.: Public Research Institute, May 1975), pp. 4-1 to 4-50.

2. The analysis is based on 1979 simply because trigger prices had their most complete effects in that year.

Figure 7-2. *The Effect of Trade Protection on U.S. Steel Prices and Output for a Domestic Oligopoly*

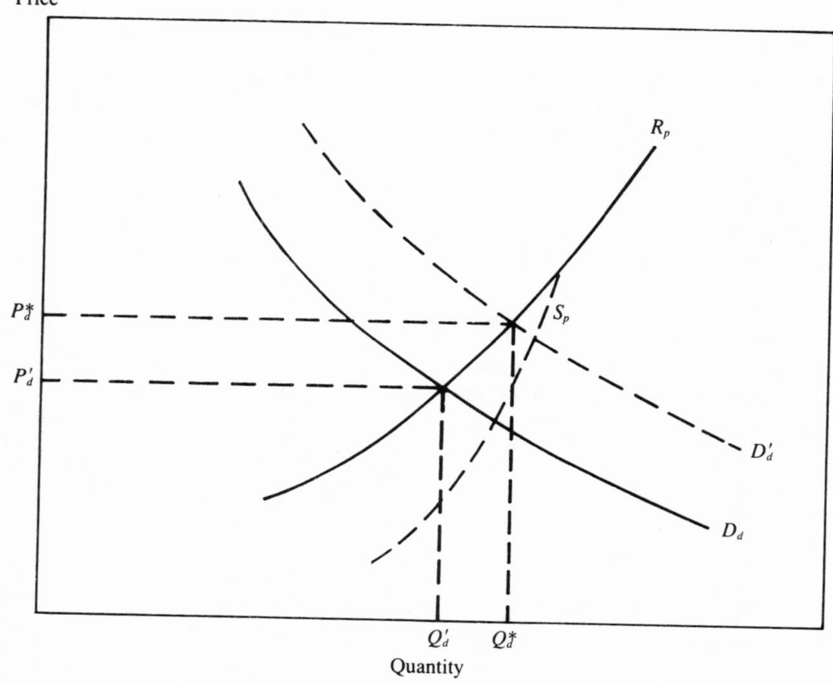

to be equal to the 1978 value, $383, escalated by the rate of increase in the producer price index for steel mill products in 1979, 10.5 percent. This results in an estimate of $423 a net ton. Imported carbon steel prices averaged $366 a net ton in 1979, to which I add $50 for importation charges, yielding a total of $416 a ton.

Distributional Effects of Trigger Prices

In chapter 5 trigger prices were found to have increased import prices by approximately 9 percent. Given estimated 1979 carbon steel import prices (including importation charges) of $416, this suggests that import prices would have been about $383 in the absence of the trigger-price mechanism. If domestic producers had responded competitively and if industry supply elasticity were 1.38, as estimated by Jondrow, one would have expected

Table 7-2. *Estimated Effects of Trigger Prices on Carbon Steel Production, 1979*

Change	*Assumed "supply" elasticity (E_s)* $E_s = 1.38$	$E_s = 3.5$
Estimated increase in import prices, including importation charges (dollars per ton)	33	33
Estimated increase in U.S. producer prices (dollars per ton)	7.20	4.20
Decrease in imports (millions of tons)	6	6.7
Increase in domestic shipments (millions of tons)	2.2	3.1
Transfer of rents to domestic resource owners (millions of dollars)	>640	>371
Additional resources attracted to domestic carbon steel industry (millions of dollars)	923	1,305
Transfer of rents to foreign resource owners (millions of dollars)	519	519
Effect on domestic employment (number employed)	+8,800	+12,400

Source: Author's estimates.

imports to rise to 23 million tons and domestic industry supply to fall to 87.8 million tons if trigger prices were abandoned. The domestic price would have fallen by $7.20 a net ton under these conditions (see table 7-2). Both effects would have required a number of years to be fully registered.

In the short run the estimates of chapter 3 suggest that the industry responds as if the elasticity of supply were 3.5 or above. Under these conditions dropping trigger prices contracts industry output to 86.9 million tons, lowers the domestic price by only $4.20 a ton, and raises carbon steel imports to 23.7 million tons.

Under the first "hypothetical" scenario, trigger prices transfer rents to domestic resources equal to $640 million a year (areas I plus III in figure 7-1). In addition $923 million in resources are attracted into the industry (areas II plus VIII). Finally, rents are transferred to foreign producers and workers in the form of higher product prices for a lower production rate. These transfers total $561 million less tariff revenues of approximately $42 million, or $519. Therefore, if displaced resources abroad find other employment at comparable returns, those remaining earn rents that compare favorably with those created for domestic labor and capital.

In fact, the industry does not respond competitively in the short run. Given an R_p curve with an elasticity of 3.5, the trigger prices have a lesser impact on prices but a greater effect on domestic output. The rents transferred to domestic suppliers of labor and capital are equal to $371 million for the 85.9 million tons of output produced without trigger prices plus a share of the $1,305 million in returns to resources attracted to the industry to produce another 3.1 million tons of steel. This latter share could be very large, particularly given the wage rates paid by the steel industry and its ability to avoid competitive price behavior. For instance, the wage bill for the 3.1 million additional tons of steel in 1979 would be about $400 million for the 12,400 additional workers. If wage rates were only 1.5 times the manufacturing industry average, the incremental wage bill would be $45 million less. Given that 1.5 times the manufacturing wage appears to be the maximum for steelworkers in other countries, this $45 million might be taken to be a reasonable estimate of the additional rents to labor accruing from the incremental output. Jacobson has estimated that displaced workers in this industry lose 21.7 percent of their earnings over six years.[3] About half this loss is due to unemployment; therefore the implicit rents for these 12,400 workers would be very close to $45 million. But the industry has been unable to earn high returns during periods of trade protection or dollar devaluation; it let labor costs accelerate dramatically during 1969–79, and the return to capital reached the manufacturing average in only one year since 1958.

If some of the rents earned through trade protection are eventually appropriated by labor, how much would the trigger price system provide to each worker once his bargaining agent had the opportunity to negotiate for it? At 8 man-hours a ton, the domestic steel industry would have employed approximately 348,000 workers to produce 86.9 million tons of carbon steel in 1979. If these workers were able to appropriate all the gains from trade protection, they would gain more than $1,000 a year per person. If they only were able to negotiate for half of this amount, they would reap $500 each.

The total rents paid to foreign resources by the trigger price system remains at $519 million under either assumption concerning domestic price response. Foreign resources are transferred out of steel (unless export prices fall sufficiently), but those that remain earn $519 million in additional rents. This is very nearly the same amount as those earned by domestic resources from the trigger price policy. Therefore, one must conclude that—assuming steel-producing resources transfer without cost from foreign industries to other

3. Louis Jacobson, "Earnings Losses of Workers Displaced from Manufacturing Industries," in U.S. Department of Labor, Bureau of International Labor Affairs, *The Impact of International Trade and Investment on Employment* (GPO, 1978).

pursuits—the trigger price mechanism has benefited foreign claimants as much as or more than the owners of domestic resources.

Effects of Protection Required to Generate New-Plant Construction

How much more would be required for the U.S. industry to begin considering new-plant construction? Surprisingly, at 1979 demand levels there is very little room for a new plant without retirements of some existing capacity. Assume that the market elasticity of demand for steel is −0.45 (Jondrow's estimate) and that imported and domestic steel are perfect substitutes. Then an 8 percent domestic price increase over 1979 levels, the minimum required for stimulating new-plant construction, would generate a 3.6 percent reduction in domestic steel consumption. Domestic carbon steel consumption (plus modest exports) would only have been 109 million net tons in 1979 under free trade conditions. An increase in imported and domestic prices through trade protection that would induce new construction would reduce consumption to 105 million tons. In 1979 the U.S. industry could probably have supplied 99 million net tons with its existing capacity. Therefore, only one new plant with a capacity of 6 million net finished tons could be built even under these most optimistic assumptions.

But domestic and imported steel are obviously not perfect substitutes. Given the results of chapter 3, an 8 percent increase in domestic prices could only be brought about through extreme trade restrictions. It is obviously unreasonable to assume that trade protection would be used to raise import prices sufficiently to give domestic producers the opportunity to raise their prices by 8 percent over the trigger-price-supported levels. Nor is it reasonable to use the results of chapter 3 to attempt to estimate the required imported price increases that would generate these domestic price increases. With very large increases in import prices and the industry nearing its capacity constraint, domestic prices would surely rise at a rate more rapid than $1 for every $10 of increase in import prices.

A more fruitful exercise would be to examine the quantitative restraints on imports that might generate sufficient domestic demand and domestic price increases to make the investment in a new plant appear profitable. If, for instance, imports were reduced to, say, 3 million tons a year, and domestic producers were allowed to increase prices by 8 percent to $457 a net ton, the effect on domestic demand would be similar to forcing import prices up to

$655 a net ton, given equations 7-1 and 7-2. This in turn would increase the demand for domestic steel to 105.2 million tons, thus requiring new capacity. While this estimate is obviously very sensitive to the assumed parameters of the demand functions, it provides some idea of just how restrictive trade policy would have to be to stimulate new plant investment in the United States.

Clearly the gainers from a policy of trade restraint of this magnitude would be domestic producers and workers. The transfers of wealth from consumers to producers would be at least $34 times the 90 million tons produced under trigger prices in 1979, or $3 billion. In addition a substantial share of the $6.3 billion in revenues obtained by the industry as it moved from 90 million tons of output to a 105-million-ton-capacity output would be rents to the steel firms and their workers. Foreign suppliers would lose about 14 million tons of output, and if the program were administered so as to force these suppliers to compete for the paltry remaining market in the United States, their prices would probably fall. The scenario is so restrictive that it is difficult to see how it could be effected without major retaliation from the supplying countries. Hence the full welfare effects of the policy can hardly be examined in a partial equilibrium framework.

Economic Welfare

It would be unfortunate to conclude a discussion of the gains and losses accruing to different groups from trade protection without at least a cursory examination of the deadweight losses from such a policy. Among the benefits of direct subsidies is the avoidance of large-scale resource misallocation. While these deadweight losses might be small relative to the multibillion dollar transfers of wealth occasioned by trade policy, they are nevertheless important.

The efficiency losses from the protection afforded by the Voluntary Restraint Agreements or the trigger prices derive mostly from resource misallocation in the production of steel. Given the limited shift of production from foreign countries to the United States, these efficiency losses are rather small. The increase in domestic carbon steel production of 3.1 million tons of finished products attracts resources to endeavors that have greater value elsewhere, but the annual total welfare loss from this source is only $22 million a year, ignoring the costs of transferring resources out of the exporters' industries.

The losses in consumer surplus are more difficult to measure, given incomplete information concerning compensated demand elasticities. The total reduction in carbon steel consumption from the trigger prices is 2 million tons, or 1.8 percent, for an average increase in imported and domestic prices of 2.2 percent. Using these averages and a linear approximation to the "market" demand curve generates a loss in consumer surplus of only $9 million. While this is a very crude approximation to the total welfare loss in steel consumption, it is so small compared with the resource transfers from consumers to producers and workers as to be rather uninteresting for policy purposes.

Similarly the welfare losses from the draconian import restrictions required to generate new-plant construction are dwarfed by the annual transfers from producers to consumers. Losses in productive efficiency from building one plant are slightly more than $200 million a year over and above the trigger price effects. Once again, using a weighted average of price increases and linearizing the implicit "market" demand relationship yields a deadweight consumption loss of $150 million a year from this more severe policy of trade restrictions.

It might be useful to compare these estimates with those that would derive from the assumption of perfect substitution in consumption between domestic and imported steel and a market elasticity of demand of −0.45. If trigger prices raise import prices by 9 percent, all steel prices would rise by this percentage. The average price of steel is $383 before trade restrictions and $416 afterward. Deadweight consumption losses would be $74 million a year. To raise both prices by another 8 percent to stimulate new investment would increase the deadweight loss by another $64 million.[4] Obviously the assumption of perfect substitution in consumption makes the policy of using import restrictions to drive up domestic prices much more efficacious, but the empirical evidence simply will not support a theory of perfect substitution.

Conclusion

It is clear that the current policy of restricting imports transfers resources from domestic consumers to producers and workers. But the surprising conclusion is that nearly as much is transferred to foreign resource owners

4. "Deadweight loss" refers to the loss in the value of social product to the economy because resources are used to produce a less desirable array of products. In this case resources transferred from fabricating steel products because of higher steel prices generate less value in other markets than they would have in the production of a variety of steel-bearing products.

as to domestic suppliers. For the 1978–79 trigger prices, the total transfer is nearly $1 billion a year, while the net employment effect is probably about 12,400 additional steelworkers in the United States. The benefits to these workers is between $45 million and $87 million a year at first. Thereafter they might be able to bargain for sharply higher wages, to be paid from the $371 million transferred from domestic consumers.

The welfare effects of raising prices to the level required to make new-plant construction profitable are much more severe. Approximately $3 billion a year would be transferred from consumers of steel to producers. The deadweight loss in production and consumption would be in excess of $350 million a year. The import market would virtually dry up as import prices were driven sharply upward to generate the necessary domestic price increase. As chapter 6 demonstrates, there are more efficient mechanisms for subsidizing domestic investment.

VIII

The Future Prospects for the U.S. Industry and U.S. Policy

This chapter looks forward tentatively to the prospects for the U.S. steel industry and for U.S. trade policy. Prospective gains and losses from trade protection are identified, and the distribution of these gains and losses among different groups is projected.

U.S. Industry Prospects

There can be little doubt that the U.S. industry is approaching or encountering a major turning point. In 1977 Bethlehem Steel and Youngstown Sheet and Tube began a process of rationalizing their facilities, eliminating older and inefficient mills and concentrating on their most profitable facilities and product lines.[1] United States Steel has begun a similar program but has closed only two steel-producing plants with less than 2 million tons of raw steel capacity.[2]

Since no new integrated mills have been constructed in the United States since the mid-1960s, integrated facilities have had to be upgraded and modernized. Several of the smaller plants of the large integrated companies may now be candidates for closure. But the number is not large, and it is unlikely to be altered much by trade policy.

Plant Closings

In the mid-1970s the American Iron and Steel Institute's estimate of production capability of the entire U.S. industry was rather steady at 155 million tons of raw steel. This capacity rose to 160 million tons in 1977, but

1. For a discussion of the plant closings that occurred in 1977 see Lee Smith, "Hard Times Come to Steeltown," *Fortune,* December 1977, pp. 87–93; "Bethlehem's Bind in Johnstown," *Business Week,* August 15, 1977, p. 40; and "A Huge Pink Slip for an Ohio City," ibid., October 3, 1977, p. 39.

2. Agis Salpukas, *New York Times,* November 28, 1979.

the plant closings since mid-1977 have more than offset some new investments at existing plants to reduce 1979 capacity to 153 million tons.[3] In 1977 four plants were either partially or totally closed:

Plant	*Capacity (millions of raw tons)*
Alan Wood Steel Co., Conshohocken, Pa.	1.1
Bethlehem Steel Corp., Johnstown, Pa.	0.6
Bethlehem Steel Corp., Lackawanna, N.Y.	2.0
Youngstown Sheet and Tube, Youngstown, Ohio	1.7
Total	5.4

In 1978 U.S. Steel announced that it had retired another 1 million tons of steelmaking capability. In 1979 U.S. Steel closed its Youngstown, Ohio, and Torrance, California, plants, idling another 1.2 million tons, and Jones and Laughlin closed its last 0.9 million tons of capacity in Youngstown. Thus in just two years the industry had retired more than 5 percent of its capacity, or 8.5 million tons.

How vulnerable are the remaining plants? No definitive analysis can be provided without detailed information on prospective pollution control costs, renovation possibilities, and information on materials and product flows, but an estimate may be made from an industry study. In 1975 A. D. Little reported the percentage of production of major carbon steel products that derived from plants with average costs 15 percent and 20 percent above the industry average.[4] For these products, which constituted 77 percent of industry shipments in 1972, 4.7 percent of production emanated from plants with costs at least 20 percent above average. Another 3.4 percent had costs at least 15 percent above average for the industry. Given noncapital and necessary refurbishing costs of $375 per net finished ton for an average carbon steel operation in 1978, this means that 4.7 percent of the industry had costs of at least $450 per net ton in 1978 and another 3.4 percent had costs of at least $431 per net finished ton. With average revenue per ton equal to $380–$385 per net finished ton of carbon steel products, at 1979 prices, obviously the least efficient 8.1 percent would not have covered costs. In fact, this 8.1 percent translates into 10.5 million tons of carbon steel capacity, and the industry's closings in 1977–79 totaled 8.5 million tons. Thus the recent

3. Information on current capacity was supplied by the American Iron and Steel Institute.
4. Arthur D. Little, Inc., *Steel and the Environment: A Cost Impact Analysis,* A report to the American Iron and Steel Institute (Cambridge, Mass.: A. D. Little, May 1975), pp. 6–52.

closings total more than 80 percent of the production identified by A. D. Little as 15 percent more costly to operate than the industry average.

Assume that the elasticity of plant costs with respect to output is equal to Jondrow's estimate of long-run supply elasticity, 1.38, for plants with costs between the industry average and 15 percent above the industry average in 1977–79.[5] This would mean that another 10.8 percent of capacity has unit operating costs at least 5 percent above the average that existed before the 1977–79 closings. These plants could conceivably be in danger if most of these costs are avoidable through plant closure. This would suggest that another 17 million tons (10.8 percent of the preclosure 159 million tons of capacity) might be candidates for closure in the next few years, particularly during recessions. The American Iron and Steel Institute, on the other hand, sees 25 to 30 million tons in danger of closing unless policies are designed to channel more capital to the industry—an estimate that seems unduly pessimistic.[6]

Recent events and an examination of industry plants would seem to confirm the deductions above.[7] National Steel derives its raw steel from three plants, each in inland locations with large, efficient basic oxygen furnaces. Bethlehem obtains approximately half of its output from Burns Harbor and Sparrows Point. The latter plant has two basic oxygen furnace vessels and the most modern large blast furnace in the country. While it is located in an exposed position in Baltimore Harbor, it is not likely to be closed any time in the foreseeable future despite its import competition problem. In addition Bethlehem has announced its intention to build electric furnaces in Johnstown to keep its 1.2 million tons of capacity there, and its Bethlehem plant appears secure. Other than the Lackawanna plant, which has cut back substantially, the remainder of Bethlehem's capacity is in modest-sized electric furnace plants that are unlikely to be threatened.

Surprisingly, the largest firm, U.S. Steel, has the greatest number of plants that would appear to be less than minimum efficient scale. The Youngstown works has been closed. Two other open-hearth facilities appear safe. The Homestead and Fairless works are reputed to be relatively efficient, and each has more than 3 million tons of open-hearth capacity. But U.S. Steel has four

5. James Jondrow, "An Econometric Model of the Steel Industry," in James Jondrow and others, "Removing Restrictions on Imports of Steel" (Arlington, Va.: Public Research Institute, May 1975), pp. 3-1 through 3-63.

6. American Iron and Steel Institute, *Steel at the Crossroads: The American Steel Industry in the 1980s* (Washington, D.C.: AISI, January 1980), p. 39.

7. The data on plant size are derived from Securities and Exchange Commission Forms 10K and *Iron and Steel Works of the World,* 7th ed. (Metal Bulletin Books, 1978).

relatively small plants, each with less than 3 million tons of basic oxygen furnace capacity. Two are located in the Monongahela Valley (Thomson-Irvin and National-Duquesne), one is in Fairfield, Alabama, and one is in Lorain, Ohio. It would not be surprising if some major rationalization plan for all of the Monongahela plants, including the Homestead, Clairton, Thomson-Irvin, and National-Duquesne plants, were to be announced by U.S. Steel within the next few years. This could involve the loss of some of the 5–6 million tons of crude steel capacity located there. The Lorain plant would appear to be much better located and therefore safe despite some materials handling problems. The Alabama plant is quite clearly threatened, as recent U.S. Steel pronouncements have indicated. Finally, the South Chicago works have been cut back recently and could be reduced further, but the giant Gary works will survive despite a variety of technical and labor problems.

A similar picture emerges when one examines the next three largest firms—Armco, Republic, and Jones and Laughlin. More than half of Armco's capacity is in Ashland, Kentucky, and Middletown, Ohio, and each of these plants appears relatively secure. The remainder is scattered among electric arc plants, mostly in Houston and Kansas City. None of these plants is likely to be imperiled by environmental policy or high costs during a recession. Republic has one small plant in the Mahoning Valley but is spending substantial sums to renovate it. Its Cleveland plant continues to rely somewhat on open-hearth furnaces, but the size, flat-rolled product mix, and location of this plant would seem to argue for its permanence. Of its remaining carbon steel facilities, the Gulf Steel Works in Gadsden, Alabama, is obviously the more poorly located and of suboptimal scale. The Buffalo works is small, but it is primarily a bar mill. Finally, Jones and Laughlin's Monongahela Valley plant in Pittsburgh has been fitted with new electric furnaces. While some question the wisdom of this investment, the furnaces are obviously modern and not likely to be closed during a recession unless electric power shortages emerge. Jones and Laughlin's remaining Youngstown capacity of 0.9 million tons has been closed, and the company could still encounter problems with its Indiana Harbor plant.

For the largest six companies, therefore, only U.S. Steel appears to have a large amount of small, old, poorly located capacity. It could be faced with decisions to close or rationalize substantial amounts of capacity in the next few years. Republic and Jones and Laughlin would appear to have perhaps another 3 million tons of capacity that could be closed in the next five or ten years. Of the remaining smaller producers, only Wheeling-Pittsburgh Steel Corporation's 2-million-ton Monessen, Pennsylvania, plant; Kaiser's Fontana,

California, works; and CF&I Steel Corporation's 1.9-million-ton plant in Colorado could conceivably be considered endangered. Kaiser has recently modernized its works in the import-plagued West Coast market, and while this might not have been a wise investment decision, this plant might survive. In short, one can identify perhaps 15–20 million tons of raw steel capacity that might be phased out in the next few years. As this capacity is closed, firms will move to expand their more efficient works, and electric-furnace producers will continue to grow. Additional problems might derive from outdated rolling facilities, but these problems cannot be assessed from public data. Nevertheless, it seems unlikely that more than 15 million tons of raw steel capacity could be lost over the next business cycle.

Locational Implications

It is not surprising that most of the potentially inefficient capacity identified here is located in one of three general areas—the Monongahela Valley, the Mahoning Valley, and the Southeast. Firms have allowed these facilities to degrade or have not upgraded them to minimum efficient scale because of the locational implications of events since 1959. Efficient integrated production in the United States requires proximity to Great Lakes ore and transportation facilities and distance from competitive imports. Thus, except for major new investments at Sparrows Point, Maryland (Bethlehem), Fontana, California (Kaiser), and Baytown, Texas (U.S. Steel), the major investments by the integrated companies have generally been in the north central part of the country, particularly on the Great Lakes. The small plants in the Pennsylvania or Ohio river valleys have generally been allowed to decay as well. The result is that four states with access to the Great Lakes—Ohio, Indiana, Illinois, and Michigan—have gradually increased their share of production to more than 50 percent of the nation's raw steel output (see table 8-1). On the other hand since 1956 the Northeast (including Pennsylvania) has lost more than one-fourth of its share of output. The West, similarly threatened by imports, has also suffered gradual decline despite the generally more rapid growth of this region of the country.

The growth of southeastern and non–Great Lakes midwestern production can be attributed almost entirely to the development of electric-furnace production. Electric-furnace production is very nearly 25 percent of U.S. output and much of this is to be found scattered around the country in small mills. This growth, at the expense of the basic oxygen furnace and the blast furnace, demonstrates the severity of the decline in the comparative advantage of the

Table 8-1. *Geographical Distribution of U.S. Raw Steel Production, 1956, 1966, 1976, and 1978*
Percent

Location	*1956*	*1966*	*1976*	*1978*
Northeast	38.7	36.0	29.2	28.2
New England, New Jersey, Delaware, Maryland	6.0	6.2	4.6	4.6
New York	5.4	5.8	3.7	3.1
Pennsylvania	27.3	24.0	20.9	20.5
Southeast[a]	7.2	8.1	9.2	9.7
Great Lakes	46.4	46.3	51.5	50.3
Ohio	19.4	17.1	17.5	15.5
Indiana	12.5	13.5	17.3	17.8
Illinois	8.4	8.2	8.6	9.1
Michigan	6.1	7.5	8.1	7.9
Midwest[b]	7.8 (Midwest and West combined)	3.4	4.0	5.7
West[c]		6.4	6.1	6.1
Total United States	100.0	100.0	100.0	100.0

Source: *Annual Statistical Report: American Iron and Steel Institute*, selected years. Figures are rounded.
a. Includes Alabama, Arkansas, Florida, Georgia, Kentucky, Louisiana, Mississippi, North Carolina, South Carolina, Tennessee, Virginia, and West Virginia.
b. Includes Iowa, Minnesota, Missouri, Nebraska, Oklahoma, and Texas.
c. Includes Arizona, California, Colorado, Hawaii, Oregon, Utah, and Washington.

U.S. industry. Its proximity to low-cost coal and iron ore is no longer a sufficient advantage to offset high labor costs.

Given the location of marginal mills in Pennsylvania, Ohio, and Alabama, and the possibility that as much as 11 million tons of additional capacity in these areas may eventually be shut down, it is obvious that the Great Lakes area will assume an even more important role in the future of the U.S. industry. Trade protection, which serves to reduce imports and raise domestic prices, cannot alter this general trend. The choice between rounding out or expanding Great Lakes plants and resuscitating marginal plants in other locations is unlikely to be very sensitive to the general level of steel prices. The optimal choice of location of production is only marginally affected by trade policy through the effect on the geographical distribution of output prices. But given the preponderance of marginal capacity in the East and Southeast, trade policy cannot possibly extend the lives of these plants very much. Trade protection will increase the value of plants such as Fairless, Sparrows Point, Baytown, and Fontana, but most of these plants are not facing the wrecking ball. Moreover, even if one believed that the price effect from trade protection were the determining factor in plant shutdown decisions, the most the 1 percent increase in prices from the trigger price mechanism could accomplish would be to rescue 3 million tons of capacity from the threat of closure. This would preserve the jobs of perhaps 12,400 steelworkers, a modest benefit for a policy of expensive trade protection.

In short, production of steel will continue to move toward the Great Lakes regardless of trade policy. No more than 10 percent of the industry's capacity appears in danger of being closed in the foreseeable future, and most of this will probably be abandoned regardless of the level of output prices. Only specific subsidies to the Monongahela and Mahoning valleys or perhaps even Alabama can prevent the gradual movement of steel capacity toward the Great Lakes.

Industry growth

Chapter 4 presented the reasons for the absence of new-plant investment in the integrated sector of the U.S. steel industry. These conditions appear unlikely to change unless there is a major change in relative prices or exchange rates. But the industry seems equally unlikely to contract in the near future since no more than 10 percent of existing capacity is likely to close because of obsolescence. These closings will be offset by additions to the efficient mills, such as Inland Steel's round out of Indiana Harbor, or potentially, expansion of Jones and Laughlin's Indiana Harbor plant or of Bethlehem's Burns Harbor facility. The addition of 11–12 million tons of capacity at existing mills can be accomplished fairly easily but at a cost of $8 billion or more.

The only major source of new growth in the U.S. industry will come from electric-furnace mini-mills. Already several of these firms are announcing important expansions of several hundred thousand tons each.[8] These producers will continue to offer principally bar products, including small structural shapes, and they will avoid the larger-scale flat-rolled products. If direct reduction from coal becomes commercially practicable, these electric-furnace operations may grow even more rapidly, but it seems unlikely that the sheet products will be produced from the direct-reduction-electric-furnace technology in the United States in the next ten or fifteen years.

There can be little doubt that the U.S. industry will not grow even if the current trade protection is extended. It would take monumental trade protection to induce new-plant investment in the blast furnace and basic oxygen furnace technology, given the comparative advantage of the less developed countries

8. Florida Steel and Nucor have each announced major expansion plans.

in this activity. Virtually every forecast of future capacity sees very little growth in the United States or the European Economic Community (EEC) but substantial growth in Latin America, the Middle East, eastern Asia, and even Africa. As recently as 1974 these areas accounted for only 8 percent of western world capacity. Central Intelligence Agency forecasts show that by 1985 these countries will account for 110 million to 115 million metric tons of capacity, or about 15 to 16 percent of non-Communist world capacity. In fact, the CIA report shows that 60 percent of the growth in capacity during the 1970s occurred in Brazil, Mexico, and India. Thus the trend toward the less developed countries is well under way.[9]

While it is difficult to be precise, the best prediction for U.S. steel capacity in the next decade is that it will be little changed from 1978 or, for that matter, from 1968. The U.S. steel industry, once a growing entity based on low-cost coal and iron ore, at best will grow very slowly through the expansion of scrap-based electric-furnace technology and modernization of a few well-located mills. At worst, its capacity will decline by 10 percent.

Whither U.S. Trade Protection?

The dynamics that forced the United States into the protective policy of 1978–79 have been described in chapter 2. The trigger price mechanism was announced as a temporary policy designed to deal with the claims of dumping under the Trade Act of 1974 amendments to the Antidumping Law. It was "suspended" in March 1980 when U.S. Steel filed a large dumping suit against seven European exporters. It has now been restored as part of a strategy to induce U.S. Steel to drop its dumping suit.

In early 1979 it would have been impossible for the Carter administration to consider abandoning the trigger price mechanism. The Multilateral Trade Negotiations were being completed and ratified by Congress. Any abandonment of the TPM, followed by dumping suits against the Europeans, would have been extremely disruptive of the negotiations. Until the negotiations were ratified in July, there was no prospect for abandoning the TPM.

It is clear that dropping the TPM would be most likely to succeed during a period of strong U.S. demand and after a year or two of a fall in the U.S. exchange rate, particularly against the yen. With U.S. demand strong, domestic mills are apt to be near capacity and unable to prove "injury" under

9. U.S. Central Intelligence Agency, National Foreign Assessment Center, "The Burgeoning LDC Steel Industry: More Problems for Major Steel Producers" (CIA, July 1979), pp. 1, 3, 4.

the Antidumping Law. In addition, if the dollar is weak, import pressures will obviously abate. But what is important for the steel trade is the relative value of the dollar lagged one or two years. As chapter 3 demonstrated, imports respond to price differentials only after a rather substantial lag. It is not this year's exchange rate but the exchange rates over the last two years that drive this year's imports. Since the U.S. dollar bottomed in late 1978, the impacts on imports in 1979 and 1980 could have been predicted. Thus late 1979 or 1980 were ideal times to drop the TPM because imports were at relatively low levels. Unfortunately the collapse of the domestic steel market in mid-1980 brought new pressures for import protection.

It is important to stress that free importation of steel into the United States is an impossibility under U.S. trade laws. Two of the major sources of imports are the aging mills of Europe and the new mills in developing countries. Imports from the former can be successfully repelled by invoking the cost-of-production standard in the Antidumping Law whenever U.S. mills are operating with excess capacity and are therefore able to plead ''injury'' to the International Trade Commission. New mills in developing countries are likely to be attacked under the countervailing duty provisions of the Trade Act of 1974, which has been strengthened procedurally in recent legislation since most of these countries offer inducements to steel investment or maintain state ownership of the facilities. Thus most exporters to the United States, with the exception of Japan, Canada, and Mexico, are candidates for either dumping or countervailing duty suits as the U.S. steel industry slips into a recession.[10]

Given the relatively mild effects of the TPM (chapter 5), it is not surprising that the Carter administration chose the politically easy course of restoring the system. Dumping and countervailing duty suits might not have had as great an effect on import prices and consumer welfare as the TPM, but they certainly would have complicated foreign economic policy. As long as the Japanese continue to have excess capacity or as long as countries such as Korea, Mexico, and Canada continue to expand their steelmaking capacity, proceedings under the U.S. trade laws are more apt to affect the distribution of steel imports rather than their volume. But this redistribution at a time when pressure exists within the Organisation for Economic Co-operation and Development (OECD) for ''rationalization'' and ''restructuring'' of the world industry could prove very difficult for officials at the Departments of Commerce and State.

Over the longer term, further protection of the American steel industry can

10. Provisions of the Antidumping Law appear in 19 U.S.C., sections 160 and following and sections 303 and following.

only place additional pressure on domestic fabricating industries. Steel-using industries have suffered even greater increases in import penetration than the steel industry itself. As table 8-2 demonstrates, imports of automobiles, railroad equipment, fasteners, fabricated structural steel, and industrial machinery have grown much more rapidly than steel imports since the 1960s. Domestic prices of steel are already at least $50 a net ton higher in the United States than in Japan. Any widening of this margin will obviously place further pressure on U.S. fabricating industries and will further slow the growth of the industrial sector of the U.S. economy.

The OECD Steel Committee—A World Cartel?

In 1977 the OECD formed a Steel Committee to attempt to resolve "structural" difficulties in the world steel industry. This committee was obviously motivated by the problems faced by the American, European, and Japanese industries in adjusting to the new realities of world steel production. The EEC clearly overexpanded its industry in the 1960s while failing to retire small, inefficient mills in such areas as the Saar valley, the Lorraine, and the United Kingdom. The U.S. industry operated at less than 80 percent of capacity in 1977. The Japanese industry, feeling protectionist pressures and extremely slack home demand, was operating at less than 70 percent of capacity. Steel consumption in Japan in 1977 was nearly 28 million tons below its 1973 peak. European consumption was 20 million tons below its peak, while U.S. consumption was off 14 million net tons. Thus these three major producing areas were consuming 62 million tons less of finished steel products in 1977 than in the frenzy of 1973.[11]

With world demand failing to recover and investment in third world countries—such as Korea, Brazil, Venezuela, and nations in the Middle East—proceeding apace, the major producers sought some means of reducing the pressures of the competition, which was escalating rapidly at the time. The first meeting of the Steel Committee was in November 1978, and predictable divisions developed.[12] Market rationalization was agreed upon as a first principle of the committee's deliberations, but actual market-sharing

11. U.S. Council on Wage and Price Stability, *Report to the President on Prices and Costs in the United States Steel Industry* (COWPS, October 1977), pp. 109, 145; and *Iron and Steel Yearbook, 1978* (Belgium: Statistical Office of the European Communities, 1978).

12. Ingo Walter, "Protection of Industries in Trouble: The Case of Iron and Steel," *The World Economy*, vol. 2 (May 1979), pp. 155–87.

Table 8-2. *U.S. Imports of Steel Fabrications and Steel Mill Products, 1965–79*
Millions of dollars

Year	Product, by Standard Industrial Classification								
	Tools and hardware (342)	*Fabricated structural metal products (344)*	*Fasteners (345)*	*Industrial machinery[a]*	*Farm machinery (352)*	*Appliances (363)*	*Autos (371)*	*Railroad equipment (374)*	*Steel (331)*
1965	71.5	16.5	51.5	571.4	221.6	105.0	917.6	6.5	1,269.0
1966	131.3	37.6	80.0	1,133.5	299.0	188.4	3,998.9	5.7	1,330.3
1967	100.6	31.3	71.1	996.9	311.9	139.5	2,480.9	5.5	1,372.6
1968	84.6	23.4	60.2	823.4	286.7	127.8	1,842.5	7.5	2,061.1
1969	152.0	48.4	99.5	1,290.0	308.4	255.7	4,931.7	5.0	1,829.5
1970	173.3	71.3	125.2	1,447.4	307.8	271.3	5,458.4	7.3	2,040.6
1971	183.5	74.4	112.1	1,568.5	325.5	325.1	7,421.9	8.0	2,738.5
1972	258.3	73.0	158.4	2,042.1	443.8	450.6	8,659.9	24.1	2,963.9
1973	295.8	76.4	233.5	2,528.2	607.7	520.5	10,117.5	31.8	3,079.6
1974	325.2	110.0	488.5	3,110.7	854.4	508.9	11,635.2	21.4	5,646.2
1975	302.0	114.0	287.1	3,481.7	1,000.6	515.9	11,426.9	52.0	4,743.2
1976	423.7	127.0	343.5	3,671.6	945.7	700.6	14,449.6	44.8	4,456.6
1977	523.8	156.7	417.1	4,517.2	1,053.8	1,443.1	17,401.8	56.0	5,969.2
1978	680.9	207.0	568.4	7,192.3	1,240.4	1,001.3	21,984.9	117.5	7,464.7
1979	732.9	213.2	605.3	8.846.1	1,782.0	973.9	23,151.6	401.1	7,566.2

Source: U.S. Bureau of the Census, ''U.S. Imports, Classified by SIC Product Code,'' Report IA-275, computer tape (n.d.).
a. All SIC 35 except 352, 357, and 358.

agreements were not. Japan and the EEC countries probably would favor some form of market sharing, the EEC as a natural extension to the trade protection already developed under the Davignon Plan (see chapter 2) and Japan as a means of guaranteeing its continued access to world markets. The United States was opposed to any such agreement.

As long as the United States, with one-fourth of the world's capacity and an import market of 20 million tons out of a world total of approximately 140 million tons, is opposed to a cartelization of the world industry, it is doubtful that a cartel will develop. The Japanese are placed in a position of negotiating with the United States and the EEC, which account for more than 25 percent of world imports, or looking for other export markets. While Japan still exports more to Asian neighbors than to the western world, it is facing the prospects of a severe contraction in these Asian markets. China and Korea account for more than half of Japan's Asian exports of steel products, and they are expanding their steel industries rapidly. Japanese exports to the United States and the EEC have been severely restricted by trade protection in those countries; hence Japan is facing severe problems in maintaining its level of steel exports.

Thus far the developing countries have shown little interest in joining the Steel Committee. Mexico, South Korea, Brazil, and India sent participants to the November 1978 meeting, but these countries and the other emerging exporters are unlikely to be interested in cartelization. With lower costs than their European and American competitors, they will expand their market shares quite naturally and reduce their dependence on imports. Only if the United States were to join in a successful developed country cartel would the emerging nations be forced to join the committee.

As long as the U.S. government succeeds in resisting the pressure from its own industry to move increasingly toward quantitative (market-sharing) agreements, there would appear to be little danger of the formation of a world cartel. The Europeans might press ahead with their own attempt to cartelize and progressively reduce non-EEC imports, but Japan, the United States, Eastern Europe, and the developing world would continue to compete in a world market constituting at least 75 percent of world steel production in the 1980s.

Summary and Conclusions

While my analysis of the changing fortunes of the U.S. steel industry does not allow me to be very optimistic about the prospects for the industry in the next few years, it does not suggest that collapse is imminent. The integrated industry must continue to retrench toward the Great Lakes while scrap-based mini-mills continue to offset a part of the decline in other areas. Once this retrenchment is complete, the industry may find itself in a stable environment of limited growth. Growth in world steel capacity will continue to occur predominantly in less developed countries. Any attempt by the United States or Europe to forestall this development through trade protection will only accelerate the decline in their steel-fabricating industries, many of which are already in trouble.

The specific conclusions that emerge from this monograph may be listed briefly:

—The decline in the position of the U.S. steel industry derives from natural economic forces of competition in a world of declining shipping costs, mobile technology, and declining real ore prices.

—The relative wages of steelworkers increased sharply in the 1970s after the devaluation of the dollar and during the Voluntary Restraint Agreements (VRAs), adding to the problems of an already declining industry. The workers' total compensation rose to 70 percent above the manufacturing average, while compensation for steelworkers in Europe remained at only 12 to 25 percent above the manufacturing average.

—Export prices and domestic steel prices each respond to changes in demand conditions as well as to changes in production costs.

—Contrary to popular belief, Japanese export prices are not sharply reduced when the Japanese steel industry has excess capacity. Rather, they respond to changes in production costs and fluctuations in world demand.

—U.S. producer prices respond to changes in world prices but not on a dollar-for-dollar basis.

—The importers' share of the domestic steel market is sensitive to changes in relative import prices but only after a substantial lag for most products.

—The U.S. industry cannot hope to modernize by building new steel mills, since the cost of producing from these mills would be higher than the cost of production in other countries, particularly Japan and some less developed countries.

—The cost of producing steel in a new plant in the United States is higher than the cost of production in most existing plants.

—The VRAs of 1969–74 increased import prices by approximately 8 percent and domestic prices by about 2 percent in 1971–72. The trigger prices of 1978–79 had a similar effect, raising import prices by approximately 10 percent and U.S. producer prices by 1 percent.

—The cost to the consumer of enforcing the antidumping laws through the trigger price mechanism of carbon steel products was nearly $1 billion in 1978, less than 2 percent of the value of steel consumed in the United States.

—Trade protection for the steel industry cannot be justified under current conditions as a means of contributing to the national security of the United States.

—Contrary to an argument often made by steel industry spokesmen, the cost of paying for imported steel in times of tight world market (shortage) conditions does not exceed the present value of the cost savings from buying low-priced imports at other times.

—Foreign producers are unlikely to form a cartel to exploit an overly dependent United States through high steel prices.

—If the United States wishes to subsidize steel capacity, it should do so directly through subsidies for new plants. The cost of such a strategy is far less than the cost of trade protection.

—The U.S. steel industry will lose capacity gradually over the next decade, but this loss will be no more than 10 percent even without trade protection.

APPENDIX A

Description of Data

This appendix presents in detail the methodology used in constructing all price and quantity series used in the empirical analysis in chapter 3.

Domestic Shipments: Products, Prices, and Quantities

The basis for the discussion of domestic steel products in chapter 3 is described in table A-1. Only finished products of carbon steel are included. The shipments quantified in table A-2 deliberately exclude transfers of steel products among plants belonging to the same company. The data describe the products, their prices (table A-3), and quantities shipped for the period 1955–1978.

U.S. Import Statistics: Products, Prices, Quantities

Data in the tables on imports describe the base used in chapter 3 in the discussion of steel products imported into the United States during the period 1956–79. Table A-4 presents a concordance of the three different classification systems used to identify imported steel products over the period studied. Tables A-5, A-6, and A-7 describe the products and their assigned valuation in the periods 1956–63, 1964–72, and 1973–79. Further data on prices and quantities are given in tables A-8, A-9, and A-10.

The data for import prices are adjusted to include duty charges, insurance payments, and transportation costs. Tariff charges for the five product categories were calculated as a weighted average of actual tariff rates. For the years 1956–62 and for the first eight months of 1963, tariff rates expressed in Bureau of the Census Schedule A categories were combined with the quantities and values of imports used to construct the basic import series to generate import charges. No listing of tariff rates in Revised Schedule A classifications is available for the years 1964–72, so tariff charges from September 1963 through 1976 were calculated using classifications from the

International Trade Commission's *Tariff Schedules of the United States, Annotated*. For this period tariff rates published in *TSUSA* were combined with quantities and values of imports expressed in *TSUSA* categories to derive import charges.

Transportation costs for the period 1956–73 were derived from the December average of the general freight index, which is available in the *Chartering Annual,* published by Maritime Research, Inc., in New York City. The index was rebased to 1 in 1966 and was then multiplied by 18.975, the average of the midpoints of ocean freight rates for steel from Europe and Japan as quoted in Richard S. Thorn, "Steel Imports, Labor Productivity, and Cost Competitiveness," *Western Economic Journal,* vol. 6 (December 1968), p. 378. This series was then divided by 1.1023 for conversion to net tons. The series, which represents the average transport costs for a net ton of steel, was multiplied by the following constants to correct for transport cost differentials for the various products:

Hot- and cold-rolled sheets	0.873
Plate	1.055
Bars and structurals	1.127

For the period 1974–77 transportation costs plus insurance were derived by taking the difference between unit CIF (cost including insurance and freight) value and unit customs value, which is available in U.S. Bureau of the Census, *U.S. Imports for Consumption and General Imports—TSUSA—Commodity by Country of Origin,* FT 246 (Government Printing Office, 1973).

One percent of customs value was then added to account for insurance costs for the period 1956–73. Insurance costs are included in the CIF valuation, so no insurance adjustment was necessary for the 1974–77 period.

Japanese Export Prices, Input Prices, and Production Costs

Tables A-11 through A-17 present the data used to compare U.S. and Japanese costs of production and to assess the relationship between these costs and Japanese export prices over the period 1956–76. Data on U.S. input prices and production costs are also presented for similar U.S. products.

It was assumed that for each of the five products, the United States and Japan employ the same technology at the efficient margin. Cost differentials are due to input price differences and different rates of growth of productivity in the two countries. Since fixed input coefficients were used, it was necessary to adjust the labor costs to account for changes in productivity.

The Japanese labor cost series was adjusted upward by 20 percent to account for fringe benefits. The labor cost, which is presented below, is the unadjusted series. It was assumed that between 1956 and 1976 Japanese productivity grew at a 10.65 percent annual rate, while U.S. productivity grew by 2.3 percent. The input coefficients for each of the five steel producers and for each of the eight inputs (iron ore, coal, electricity, gas, oil, labor, scrap, tin) were estimated by the author and represent the amount of each factor required to produce a metric ton of steel. These coefficients were multiplied by the corresponding costs, and the resulting products were summed over all inputs. This summation was then divided by 1.1023 for conversion to short tons.

Both quantities and values were taken from total Japanese exports, not merely exports to the United States. The original quantities were in metric tons and were converted to short tons to be consistent with domestic shipment and import data. Values were f.o.b. and were published in yen. the *International Financial Statistics*' ''par rate/market rate'' exchange rate series was used to convert to dollars.

Table A-11 presents a list of the products used to construct the production cost series. Table A-12 shows Japanese export prices. Tables A-13 and A-14 list the prices of inputs for Japanese and U.S. steel production, and tables A-15 and A-16 show my estimates of production costs in the five product classes over the period 1956–76. Table A-17 gives basic-materials costs in Japan and the United States.

Table A-1. *Description of Domestic Carbon Steel Products Shipped in the United States, 1955–78*

Product and period	Standard industrial classification number	Product description
Bars		
1955–60	n.a.	Hot-rolled bars, except concrete reinforcing
1961–70	3312421	Hollow drill steel
	3312423	All other bars
1971–78	3312422	Hot-rolled bars, except concrete reinforcing
Cold-rolled sheets		
1955–60	n.a.	Cold-rolled sheets
1961–78	3316711	Cold-rolled sheets
Hot-rolled sheets		
1955–60	n.a.	Hot-rolled and enameling sheets
1961–70	3312311,15	Hot-rolled and enameling sheets
1971	3312312	Hot-rolled sheets (including enameling sheets)
1972–78	3312311	Hot-rolled sheets (including enameling sheets)
Plate		
1955–60	n.a.	Floor plate
	. . .	Plate other than floor plate
1961–78	3312411	Floor plate
	3312413	Plate other than floor plate
Structurals		
1955–57	n.a.	Structural shapes and piling
	. . .	Piling
1958–60	n.a.	Structural shapes and piling
1961–70	3312415	Structural shapes
	3312417	Sheet piling
	3312419	Bearing piles
1971	3312416	Structural shapes (heavy), sheet piling, and bearing piles
1972–78	3312415	Structural shapes (heavy), sheet piling, and bearing piles

Sources: Data for 1955–58 are from U.S. Department of Commerce, Bureau of the Census, *Facts for Industry: Steel Mill Products, 1955*, series M22B-04 (Government Printing Office, 1957), and subsequent issues; later data are from U.S. Department of Commerce, Bureau of the Census, *Current Industrial Reports: Steel Mill Products*, series MA-33B (GPO, 1959), and subsequent issues through 1978.

n.a. = Not applicable.

Table A-2. *Domestic Shipments of Carbon Steel Products, 1955–78*
Millions of short tons

Year	*Bars*	*Cold-rolled sheet*	*Hot-rolled sheet*	*Plate*	*Structurals*
1955	6.43	13.79	9.58	6.44	5.06
1956	6.78	12.32	8.66	7.47	5.50
1957	6.00	10.94	7.70	9.12	7.00
1958	4.48	9.53	6.18	5.19	4.11
1959	5.18	11.72	7.78	5.46	4.39
1960	5.46	13.47	8.05	5.82	5.20
1961	5.03	11.26	7.39	5.29	4.49
1962	5.55	12.34	8.34	5.56	4.49
1963	5.88	13.11	9.31	6.11	4.91
1964	6.60	14.34	10.74	7.03	5.49
1965	7.27	15.01	11.44	7.65	6.09
1966	7.10	14.38	10.69	6.96	6.08
1967	6.20	13.28	10.03	6.10	5.42
1968	6.79	14.53	11.40	6.46	5.44
1969	6.98	14.87	13.57	6.21	5.53
1970	6.17	12.68	13.46	6.21	5.42
1971	5.55	13.31	11.83	6.03	5.17
1972	6.38	14.10	14.56	5.85	5.00
1973	7.67	18.27	16.42	7.28	6.24
1974	7.73	16.46	16.30	8.30	6.46
1975	5.75	11.81	12.41	6.64	4.52
1976	6.24	16.48	15.84	5.36	3.75
1977	6.49	15.34	15.07	5.63	3.98
1978	7.20	15.59	15.74	6.20	4.50

Sources: Data for 1955–58 are from U.S. Department of Commerce, Bureau of the Census, *Facts for Industry: Steel Mill Products, 1955,* series M22B-04 (GPO, 1957), and subsequent issues; later data are from U.S. Department of Commerce, Bureau of the Census, *Current Industrial Reports: Steel Mill Products,* series MA-33B (GPO, 1959), and subsequent issues through 1978.

Table A-3. *Average Realized U.S. Producers' Prices of Carbon Steel Products, 1956–78*
Dollars per short ton

Year	*Bars*	*Cold-rolled sheet*	*Hot-rolled sheet*	*Plate*	*Structurals*
1956	123.73	125.05	110.56	114.33	110.79
1957	135.65	135.01	118.52	124.88	121.48
1958	142.80	138.51	121.15	129.35	125.35
1959	147.44	140.96	122.64	126.67	128.12
1960	144.76	139.71	122.77	133.49	127.84
1961	145.19	137.83	121.14	132.23	127.88
1962	143.64	137.76	117.89	131.92	128.17
1963	144.37	140.20	119.04	131.25	127.31
1964	147.19	141.18	118.72	134.53	130.14
1965	152.73	141.66	120.86	134.71	130.88
1966	155.04	142.08	120.40	136.48	132.31
1967	152.01	142.93	120.39	138.32	133.54
1968	152.92	146.53	120.27	145.23	134.64
1969	144.60	152.43	119.89	152.06	141.66
1970	166.39	160.64	128.25	162.12	153.28
1971	179.25	166.39	136.81	175.05	166.19
1972	192.80	178.15	142.99	187.64	175.20
1973	204.96	185.57	153.72	199.86	184.43
1974	269.35	248.42	206.18	252.07	235.55
1975	294.10	269.02	221.67	292.83	274.29
1976	310.74	288.13	236.89	302.66	284.40
1977	335.63	322.13	257.38	323.59	289.63
1978	365.22	354.88	286.68	357.01	324.69

Sources: Data for 1955–58 are from U.S. Department of Commerce, Bureau of the Census, *Facts for Industry: Steel Mill Products, 1955*, series M22B-04 (GPO, 1957), and subsequent issues; later data are from U.S. Department of Commerce, Bureau of the Census, *Current Industrial Reports: Steel Mill Products*, series MA-33B (GPO, 1959), and subsequent issues through 1978.

Table A-4. *Product Codes Used to Identify Imported Carbon Steel Products Included in the Price and Quantity Series*[a]

	Product code		
Product class	*1956–63*[b]	*1964–72*[c]	*1973–79*[d]
Bars	6008000,8100,8200 6008300,8400,8500 6008600,8700,8800	6732420	608.4520,40,60 608.4620,40,60
Cold-rolled sheets	6056730,6056830 6057300,6057430	6744130	608.8744
Hot-rolled sheets	6056200,700,800 6057200,6000,6500 6056600,6038100	674415 6744125 6744120[e]	608.8440 608.8742
Plate	6038000,200,300	6741520	608.8420 608.8410,15[f]
	6038400,500 6039700 6038230,6038430	6741550 6741540[g]	608.8720
Structurals	6081020,40 6081060,1100 6081300	6734020 6734060	609.8005,15,35,41,45 609.8200,.8800,.9000 609.9600

a. Imports for consumption only.

b. The classification system for this period is the old Schedule A. See U.S. Bureau of the Census, *United States Imports of Merchandise for Consumption—Commodity by Country of Origin,* FT 110 (Government Printing Office, 1956), and subsequent annual issues through 1963. In 1963 the Schedule A numbers covered only the months of January through August. To include imports in the months of September through December, it was necessary to use classification data in International Trade Commission, *Tariff Schedules of the United States, Annotated.* The *TSUSA* categories were as follows: bars, 608.4500, .4600; cold-rolled sheets, 608.8740; hot-rolled sheets, 608.8440; plate, 608.8420, .8720; structurals, 609.8020, .8200, .8800, .9600.

c. The classification for this period is the revised Schedule A. See U.S. Bureau of the Census, *U.S. Imports of Merchandise for Consumption,* FT 125 (GPO, December 1964), and December issues for 1965 and 1966; and U.S. Bureau of the Census, *U.S. Imports, General and Consumption—Schedule A—Commodity by Country,* FT 135 (GPO, December 1967), and subsequent December issues through 1972.

d. The classification system for this period is *TSUSA*. See U.S. Bureau of the Census, *U.S. Imports for Consumption and General Imports—TSUSA—Commodity by Country of Origin,* FT 246 (GPO, 1973), and subsequent annual issues through 1979.

e. 1964–67.

f. 1977–79.

g. 1964 and 1965 only.

Table A-5. *Description of Imported Carbon Steel Products, 1956–63*

Product, by Schedule A classification number	*Product description*[a]
Bars	
6008000	Steel bars, not elsewhere specified, not over 1.5 cents a pound
6008100	Steel bars, 1.5–2.5 cents a pound, not over 30 pounds a foot
6008200	Steel bars, 1.5–2.5 cents a pound, over 30 pounds a foot
6008300	Steel bars, 2.5–3.5 cents a pound
6008400	Steel bars, 3.5–5 cents a pound
6008500	Steel bars, 5–8 cents a pound
6008600	Steel bars, 8–12 cents a pound
6008700	Steel bars, 12–16 cents a pound
6008800	Steel bars over 16 cents a pound
Cold-rolled sheets	
6056730	Steel sheets, cold rolled 1/100–109/1,000 inch, over 3 cents a pound
6056830	Steel sheets, cold rolled 1/100 inch thick, over 3 cents a pound
6057300	Steel sheets and plates, not specified, 5 to 8 cents a pound
6057430	Steel sheets and plates, not specified, cold rolled, 8 to 12 cents a pound
Hot-rolled sheets	
6056200	Steel sheets, 10/1,000–22/1,000 inch thick, not over 3 cents a pound
6056700	Steel sheets, 1/100–109/1,000 inch thick, over 3 cents a pound
6056800	Steel sheets under 1/100 inch thick, over 3 cents a pound
6057200	Steel sheets and plates, not specified, 2.5–5 cents a pound
6056000	Steel sheets, 38/1,000–109/1,000 inch thick, not over 3 cents a pound
6056500	Steel sheets, corrugated or crimped, over 3 cents a pound
6056600	Steel skelp, over 3 cents a pound
6038100	Skelp, over 3 cents a pound, over 109/1,000 inch thick
Plate	
6038000	Boiler or other plate iron, skelp, not over 3 cents a pound
6038200	Steel plates over 3 cents a pound, not over 48 inches wide, 109/1,000–23/100 inch thick
6038300	Steel plates over 3 cents a pound, not over 48 inches wide, not under 23/100 inch thick
6038400	Steel plates over 3 cents a pound, over 48 inches wide, 109/1,000–18/100 inch thick
6038500	Steel plates over 3 cents a pound, over 48 inches wide, not under 18/100 inch thick
6039700	Sheets or plates of iron or steel, polished, etc.
6038230	Steel plate, cold rolled, over 3 cents a pound, not over 48 inches wide, 109–230/1,000 inch thick
6038430	Steel plate, cold rolled, over 3 cents a pound, over 48 inches wide, 109–180/1,000 inch thick

Table A-5. *(Continued)*

Product, by Schedule A classification number	*Product description*[a]
Structurals	
6081020	Steel beams, etc., not assembled, over 3 inches wide
6081040	Steel beams, etc., less than 3 inches wide
6081060	Steel structural shapes, machined, etc., not elsewhere specified
6081100	Steel beams, etc., machined, etc.
6081300	Iron and steel sheet piling

Sources: U.S. Bureau of the Census, *United States Imports of Merchandise for Consumption—Commodity by Country of Origin,* FT 110 (Government Printing Office, 1956).

a. Valuation is on a customs basis.

Table A-6. *Description of Imported Carbon Steel Products, 1964–72*

Product, by revised Schedule A classification code	*Product description*
Bars	
6732420	Steel bars, not alloyed, not coated or plated, and not cold formed
Cold-rolled sheets	
6744130	Steel sheets, cold rolled, and/or pickled, not alloyed, not shaped, and not coated
Hot-rolled sheets	
6744115	Steel sheets, not alloyed, not pickled, not shaped, and not coated
6744125	Steel sheets, not alloyed, pickled, but not cold rolled, not shaped, and not coated
6744120	Steel sheets, hot rolled, not alloyed, pickled, not shaped, and not coated (exists only in 1964–67)
Plate	
6741520	Plates of unalloyed iron or steel, not shaped, not coated, and not pickled or cold rolled
6741540	Plates of unalloyed iron or steel, not elsewhere specified (1964, 1965 only)
6741550	Plates of unalloyed iron or steel, not elsewhere specified, not coated
Structurals	
6734020	Angles, shapes, and sections of iron or steel, 3 inch or more, not advanced
6734060	Sheet piling of iron or steel

Sources: U.S. Bureau of the Census, *U.S. Imports of Merchandise for Consumption,* FT 125 (Government Printing Office, December 1964), and subsequent December issues through 1966; and U.S. Bureau of the Census, *U.S. Imports, General and Consumption—Schedule A—Commodity by Country,* FT 135 (GPO, December 1967), and subsequent December issues through 1972.

Table A-7. *Description of Imported Carbon Steel Products, 1973–79*

Product class, TSUSA *classification*	*Product description*
Bars	
608-4520	Flat steel bars, not alloyed, not coated, not cold formed, not over 5 cents a pound
608-4540	Round steel bars, not alloyed, not coated, not cold formed, not over 5 cents a pound
608-4560	Steel bars, not elsewhere specified, not coated, not cold formed, not over 5 cents a pound
608-4620	Flat steel bars, not alloyed, not coated, not cold formed, over 5 cents a pound
608-4640	Round steel bars, not alloyed, not coated, not cold formed, over 5 cents a pound
608-4660	Steel bars, not elsewhere specified, not alloyed, not coated, not cold formed, over 5 cents a pound
Cold-rolled sheet	
608-8744	Steel sheet pickled and/or cold rolled, not shaped, not alloyed
Hot-rolled sheet	
608-8440	Steel sheet, not shaped, etc.; not cold rolled, not plated, not alloyed
608-8742	Steel sheet, pickled and hot rolled, not shaped, not alloyed
Plate	
608-8420	Steel plate, not shaped, not cold rolled, etc.; not coated, not alloyed
608-8720	Steel plate, cold rolled, etc.; not shaped, not alloyed
608-8410	Steel plate in coils, not shaped, not cold rolled, etc.; not alloyed
608-8415	Steel plate not in coils, shaped, cold rolled, etc.; not alloyed
Structurals	
609-8005	Steel, wide flange H-piles, not alloyed, not advanced, 3 inches and over
609-8015	Steel, wide flange shapes or sections, not alloyed, not advanced, 3 inches and over
609-8035	Steel angles, not alloyed, not advanced, 3 inches and over
609-8041	Steel channels, not alloyed, not advanced, 3 inches and over
609-8045	Steel shapes and sections, not elsewhere specified, not alloyed, not advanced, 3 inches and over
609-8200	Structural steel not advanced, alloyed
609-8800	Angles, shapes, etc.; unalloyed, cold formed, advanced, not over 0.28 pounds per foot
609-9000	Angles, shapes, etc., alloyed, cold formed, advanced, not over 0.28 pounds per foot
609-9600	Iron and steel sheet piling, not alloyed

Source: U.S. Bureau of the Census, *U.S. Imports for Consumption and General Imports, TSUSA, Commodity by Country of Origin,* FT 246 (GPO, 1973), and subsequent annual issues through 1979. Titles may vary slightly from year to year.

Table A-8. *Customs Value of Carbon Steel Imports, 1956–79*
Dollars per short ton

Year	*Bars*	*Cold-rolled sheet*	*Hot-rolled sheet*	*Plate*	*Structurals*
1956	104.91	143.78	111.25	134.83	123.60
1957	110.09	137.41	138.74	164.86	138.91
1958	77.01	116.60	102.13	105.20	114.25
1959	89.33	183.86	112.67	103.10	102.57
1960	103.27	180.57	128.22	111.78	119.29
1961	90.84	126.33	102.03	116.45	107.51
1962	86.18	121.87	95.23	122.71	104.00
1963	82.02	111.29	87.07	111.74	91.89
1964	87.50	110.21	85.54	96.45	89.81
1965	90.58	109.27	85.08	97.38	94.14
1966	88.40	104.25	82.40	89.78	92.78
1967	86.10	101.43	81.85	89.45	93.49
1968	85.88	105.22	79.72	86.58	90.94
1969	102.00	107.92	83.49	94.81	98.90
1970	113.99	126.37	102.92	119.45	119.47
1971	116.93	132.12	109.92	120.91	120.77
1972	129.70	143.78	121.67	133.31	127.40
1973	159.57	164.04	137.37	151.10	148.73
1974	293.31	281.02	232.49	279.08	275.53
1975	281.05	237.62	213.68	269.62	267.02
1976	222.03	233.30	194.94	210.81	210.92
1977	219.79	254.04	204.86	214.25	212.61
1978	261.25	285.52	233.23	236.42	258.20
1979	326.31	336.00	278.31	294.54	320.12

Sources: Same as for table A-4.

Table A-9. *Prices of U.S. Imports of Carbon Steel Products, Including Importation Charges, 1956–76*
Dollars per short ton

Year	*Bars*	*Cold-rolled sheet*	*Hot-rolled sheet*	*Plate*	*Structurals*
1956	159.34	192.83	156.58	189.63	171.08
1957	143.04	169.15	169.59	201.78	165.08
1958	103.85	145.42	128.29	134.80	138.90
1959	118.31	219.80	140.64	133.27	126.94
1960	133.53	215.75	157.41	142.23	143.72
1961	118.52	156.12	127.96	146.01	130.43
1962	111.12	148.32	118.35	149.40	124.83
1963	111.51	140.29	112.61	142.15	117.42
1964	115.47	138.08	109.39	124.42	113.30
1965	119.85	137.79	109.75	126.31	118.62
1966	115.56	130.67	105.25	116.06	115.21
1967	114.13	128.47	105.49	116.77	117.11
1968	112.79	131.40	102.23	112.22	113.38
1969	131.09	133.96	106.31	121.41	121.71
1970	146.75	156.09	129.50	150.82	145.37
1971	144.94	158.06	133.46	148.22	142.16
1972	160.47	172.52	148.02	163.74	151.13
1973	203.97	203.31	173.80	193.56	183.95
1974	333.43	310.97	256.96	311.45	307.96
1975	343.37	298.16	252.01	316.10	302.49
1976	266.86	268.28	224.00	242.82	241.33

Sources: Same as for table A-4. Tariff charges based also on data in Bureau of the Census, *Schedule A, Statistical Classification of Commodities Imported into the United States, with Rates of Duty and Tariff Paragraphs* (Government Printing Office, 1954), and ibid., 1954 ed., reprinted and updated in 1957; and ibid., 1960 ed. (titles vary slightly).

Table A-10. *Carbon Steel Imports, 1956–79*
Thousands of short tons

Year	*Bars*	*Cold-rolled sheet*	*Hot-rolled sheet*	*Plate*	*Structurals*
1956	44.6	2.7	4.0	62.1	603.6
1957	23.8	1.8	0.4	29.3	428.9
1958	78.3	0.2	6.1	27.2	299.5
1959	195.5	28.8	68.4	363.0	863.3
1960	112.3	65.8	127.3	281.6	594.0
1961	103.1	5.7	10.6	68.2	551.3
1962	113.4	54.1	51.2	204.9	684.0
1963	192.4	192.4	230.0	352.1	777.6
1964	332.4	383.6	521.9	460.3	638.2
1965	497.0	1,218.4	1,797.0	771.8	928.8
1966	500.1	1,119.7	1,946.2	943.7	946.8
1967	552.7	1,369.9	2,261.4	1,012.8	1,063.4
1968	804.9	2,825.9	3,436.5	1,755.9	1,512.7
1969	685.8	1,907.5	1,930.5	1,171.1	1,434.1
1970	527.7	2,174.9	1,997.6	944.7	1,186.2
1971	798.1	3,544.3	2,659.7	1,538.7	1,572.0
1972	791.9	3,236.2	2,231.0	1,648.3	1,712.3
1973	707.6	2,704.4	1,786.5	1,319.8	1,338.1
1974	677.5	2,547.6	1,765.0	1,693.7	1,234.7
1975	418.6	2,067.1	1,509.2	1,353.0	876.3
1976	369.1	2,350.7	1,635.9	1,555.4	1,425.0
1977	693.6	3,345.1	2,675.0	2,108.3	1,817.0
1978	587.2	3,123.4	2,613.0	2,875.0	1,926.2
1979	438.4	2,322.3	2,153.4	1,766.9	1,985.1

Sources: Same as for table A-4.

Table A-11. *Description of Products Used to Construct Japanese Cost of Production and Export Price Data*

Product class	Classification code[a]			Product description
	1956–61	1962–75	1976	
Bars	681-0411	673-221	73.10-231	Small-section round bars
	681-0412	673-222	73.10-232	Medium-section round bars
	681-0413	673-223	73.10-233	Large-section round bars
	681-0414	673-224	73.10-234	De-formed steel bars
	681-0415	673-225		Square steel bars
	681-0416	673-226	73.10-235,6,7	Flat steel bars
	681-0418	673-227	73.10-220	Mechanical structural bars
	681-0419		73.10-239	Steel bars, not elsewhere specified
Cold-rolled sheets	681-0521	674-321	73.13-321,9	Less than 0.9 millimeter thick
	681-0522	674-322	73.13-331,9	0.9–3 millimeters thick
	681-0622	675-042	73.12-042	Greater than 50 and less than 500 millimeters wide
	681-0623			Greater than 500 millimeters wide
Hot-rolled sheets	681-0511,2	674-311,2	73.13-311,9	Less than 3 millimeters thick
	681-0612	675-032	73.12-032	Greater than 50 and less than 500 millimeters wide
	681-0613			Greater than 500 millimeters wide
		672-710	73.08-000	Coils
Plate	681-0513	674-230,-131	73.13-200-120	3–6 millimeters thick
	681-0514	674-132	73.13-110	Greater than 6 millimeters thick
		674-140	73.09-000	Universal plates
		674-150		Heavy plates
Structurals	681-0421	673-412,-511	73.11-311,-321	Angles
	681-0422,3,4	673-411,3,9	73.11-100,-319	Channels and sections
		673-519	73.11-211,-212, -213,-220, -329,-500	Channels and sections
	681-0427	673-420	73.11-600	Sheet piling
	681-0426,9			Other structurals

Sources: For 1956–64 see Japan Tariff Association, *Annual Return of the Foreign Trade of Japan* (Tokyo: JTA); for 1965–76 see another annual publication of the Japan Tariff Association, *Japan Exports and Imports: Country by Commodity*.

a. Classification numbers are based on Standard International Trade Classification codes.

Table A-12. *Average Realized Japanese Export Prices for Carbon Steel Products, 1956–76*
Dollars per short ton

Year	*Bars*	*Cold-rolled sheet*	*Hot-rolled sheet*	*Plate*	*Structurals*
1956	113.17	157.14	140.69	152.97	122.73
1957	125.60	166.37	165.60	174.93	139.26
1958	81.37	147.08	126.79	110.55	120.11
1959	88.05	144.42	119.80	96.85	103.95
1960	98.27	160.14	116.1	109.17	110.74
1961	94.37	133.73	114.30	105.02	111.95
1962	81.24	117.74	81.94	90.94	94.41
1963	83.60	115.28	79.83	83.43	91.12
1964	84.36	116.58	82.32	92.80	92.66
1965	89.25	110.41	82.66	94.80	93.99
1966	84.07	106.35	79.50	86.93	94.25
1967	97.82	109.30	80.17	90.12	103.91
1968	90.55	105.37	77.24	82.66	97.52
1969	93.43	111.82	81.08	90.07	102.74
1970	115.63	132.36	95.73	110.19	124.20
1971	102.91	123.92	92.56	103.84	110.58
1972	113.96	137.84	103.49	115.35	117.77
1973	168.29	179.02	135.75	146.51	162.81
1974	275.26	263.42	212.14	256.04	273.27
1975	190.24	227.13	172.86	232.62	211.17
1976	171.79	231.92	181.33	196.28	193.14

Sources: For 1956–64 see Japan Tariff Association, *Annual Return of the Foreign Trade of Japan* (Tokyo, JTA); for 1965–76 see another annual publication of the Japan Tariff Association, *Japan Exports and Imports: Country by Commodity*.

Table A-13. *Japanese Input Prices*
Current dollars

Year	*Coal (net ton)*	*Electricity (millions of kilowatt hours)*	*Gas[a] (thousands of cubic feet)*	*Iron ore (net ton)*	*Labor (man-hour)*	*Oil (net ton)*	*Scrap (net ton)*
1956	22.14	9.07	0.49	16.69	0.43	18.10	66.00
1957	26.22	9.07	0.74	19.69	0.47	27.20	76.02
1958	19.31	9.07	0.47	14.70	0.48	17.42	43.16
1959	16.33	9.07	0.38	12.70	0.50	14.08	46.76
1960	15.63	9.35	0.44	12.88	0.53	16.12	45.56
1961	15.50	9.66	0.41	12.88	0.58	15.22	49.97
1962	15.35	10.05	0.36	12.97	0.62	13.13	37.52
1963	14.74	10.24	0.35	12.32	0.66	12.75	38.78
1964	14.43	10.24	0.33	12.21	0.75	11.98	41.16
1965	14.27	10.24	0.33	12.17	0.82	12.27	41.80
1966	14.41	10.24	0.32	11.91	0.91	11.93	40.19
1967	14.22	10.24	0.37	11.49	1.04	13.45	41.72
1968	14.40	10.24	0.35	11.10	1.17	13.07	35.43
1969	14.82	10.24	0.30	10.56	1.40	11.06	40.81
1970	18.29	10.24	0.33	10.74	1.69	12.34	50.26
1971	19.41	10.61	0.41	10.51	1.98	15.15	35.44
1972	19.87	12.07	0.42	10.37	2.48	15.30	40.08
1973	21.61	14.28	0.64	11.12	3.42	23.70	77.28
1974	40.71	24.37	1.88	13.26	4.24	69.27	121.29
1975	50.82	26.65	2.10	15.15	4.94	77.51	80.50
1976	53.60	30.63	1.75	15.81	5.25	64.48	81.82

Source: U.S. Federal Trade Commission, *The United States Steel Industry and Its International Rivals: Trends and Factors Determining International Competitiveness* (Government Printing Office, 1978), table 3.3, pp. 117–18.

a. Derived from the Japanese oil price series by dividing by 40.6 for conversion from metric tons to thousand cubic feet equivalence.

Table A-14. *U.S. Input Prices*
Dollars

Year	*Coal (net ton)*	*Electricity (thousands of kilowatt hours)*	*Gas (thousands of cubic feet)*	*Iron ore (net ton)*	*Labor (man-hours)*	*Oil (net ton)*	*Scrap (net ton)*
1956	9.85	12.58	0.41	9.63	3.35	18.52	48.04
1957	10.77	12.65	0.42	10.42	3.60	20.25	41.73
1958	10.48	13.43	0.44	10.61	3.87	16.86	34.00
1959	10.50	13.49	0.46	10.80	4.14	17.10	36.52
1960	10.56	13.69	0.47	11.15	4.19	17.46	28.76
1961	9.83	13.78	0.47	11.78	4.36	17.74	32.39
1962	9.70	13.96	0.48	11.60	4.51	17.77	25.20
1963	9.35	13.85	0.48	11.67	4.60	17.46	24.20
1964	9.85	13.72	0.47	11.88	4.63	16.86	30.06
1965	9.65	13.72	0.47	11.80	4.72	16.66	30.66
1966	9.82	13.76	0.48	11.74	4.93	16.59	27.56
1967	10.33	13.69	0.47	11.91	5.11	16.66	24.66
1968	10.59	13.78	0.49	12.31	5.37	17.10	23.08
1969	10.76	13.96	0.49	12.42	5.80	16.96	27.56
1970	12.27	14.34	0.53	13.05	6.10	21.03	36.65
1971	15.27	15.50	0.57	14.12	6.67	26.33	30.44
1972	17.68	16.36	0.62	15.04	7.46	26.06	32.93
1973	19.79	17.11	0.67	15.48	8.02	29.08	51.73
1974	34.22	21.21	0.81	19.62	9.35	68.00	96.87
1975	52.66	26.39	1.13	23.99	11.03	65.21	64.03
1976	56.04	28.36	1.24	27.62	12.14	62.49	69.45

Source: U.S. Federal Trade Commission, *The United States Steel Industry and Its International Rivals: Trends and Factors Determining International Competitiveness* (Government Printing Office, 1978), table 3.3, pp. 117–118.

Table A-15. *Japanese Production Costs, 1956–77*
Dollars per short ton

Year	*Hot-rolled sheet*	*Cold-rolled sheet*	*Bars*	*Structurals*	*Plate*
1956	75.04	82.73	77.20	77.20	87.79
1957	84.64	92.62	86.62	86.62	98.48
1958	66.97	74.42	67.46	67.46	78.49
1959	59.43	66.24	60.84	60.84	70.08
1960	58.79	65.54	60.00	59.99	69.34
1961	58.47	65.09	60.10	60.10	68.86
1962	57.62	64.28	58.12	58.12	67.29
1963	55.29	61.73	56.02	56.02	64.63
1964	55.20	61.67	56.25	56.25	64.67
1965	54.83	61.24	55.93	55.93	64.25
1966	54.41	60.83	55.40	55.40	63.79
1967	54.22	60.80	55.43	55.43	64.07
1968	53.78	60.52	54.49	54.49	63.63
1969	54.26	61.16	55.68	55.68	64.55
1970	59.06	66.33	61.13	61.13	70.38
1971	60.98	69.05	61.63	61.63	73.02
1972	64.38	73.34	65.57	65.57	77.81
1973	75.18	85.80	79.58	79.58	92.94
1974	105.98	120.74	111.32	111.32	131.91
1975	120.01	136.95	120.38	120.38	147.20
1976	121.10	137.73	121.51	121.51	145.87

Source: Author's calculations.

Table A-16. *U.S. Production Costs*
Dollars per short ton

Year	*Bars*	*Cold-rolled sheet*	*Hot-rolled sheet*	*Plate*	*Structurals*
1956	63.64	70.59	60.67	76.35	63.64
1957	66.65	74.88	64.39	80.68	66.65
1958	67.75	77.36	66.15	83.15	67.75
1959	69.97	79.70	68.11	85.90	69.97
1960	69.52	80.18	68.48	86.04	69.52
1961	71.13	81.60	69.78	87.64	71.13
1962	70.47	81.82	69.74	87.73	70.47
1963	70.09	81.50	69.45	87.37	70.09
1964	70.68	81.35	69.59	87.27	70.68
1965	70.35	80.89	69.17	86.82	70.35
1966	70.80	81.87	69.89	87.85	70.80
1967	71.58	83.09	70.96	89.03	71.58
1968	73.35	85.41	72.93	91.49	73.35
1969	76.23	88.27	75.32	94.83	76.23
1970	80.69	92.38	79.16	99.44	80.69
1971	87.27	100.99	86.63	108.18	87.26
1972	95.06	109.97	94.37	117.90	95.06
1973	121.91	115.76	99.66	124.69	101.91
1974	133.02	146.55	128.33	157.42	133.02
1975	159.86	182.92	160.17	193.89	159.86
1976	174.43	199.52	174.97	211.34	174.43

Source: Author's calculations.

Table A-17. *Basic-Materials Costs in Japan and the United States, 1956–76*
Dollars per net ton

	Basic-materials cost	
Year	*United States*	*Japan*
1956	50.89	84.52
1957	45.18	96.54
1958	47.26	62.17
1959	42.91	58.99
1960	43.91	56.31
1961	45.45	63.19
1962	42.91	52.13
1963	42.09	50.14
1964	43.55	49.20
1965	43.45	49.23
1966	42.73	46.43
1967	43.36	45.00
1968	44.45	42.60
1969	45.45	44.21
1970	51.17	49.74
1971	55.34	48.35
1972	59.51	46.80
1973	67.04	59.56
1974	104.06	94.98
1975	124.65	99.18
1976	137.08	101.87

Source: Based on data in U.S. Federal Trade Commission, *The United States Steel Industry and Its International Rivals: Trends and Factors Determining International Competitiveness* (Government Printing Office, 1978), table 3-1, p. 113.

APPENDIX B

Additional Regression Results

Table B-1. *Estimates of the Determinants of Japanese Export Prices (Equation 3-1), 1962–76*

Product	*Constant*	*JMATCOST*	*JLABORCOST*	*g*	*D74*	*$DEVTREND_t$*	*$DEVTREND_{t-1}$*	*$CUJAP_t$*	$\bar{R}^2$	ρ^a	*Durbin-Watson statistic*
Hot-rolled sheet	−29.14	0.8475	250.6	0.11	32.53	28.30	. . .	0.07154	0.997	−0.753	2.004
		(10.91)	(20.00)		(8.24)	(1.33)		(0.29)			
Hot-rolled sheet	−22.69	0.8448	249.5	0.11	32.68	33.32	. . .	. . .	0.997	−0.748	2.088
		(11.55)	(21.86)		(8.79)	(2.85)					
Hot-rolled sheet	−51.99	0.8164	257.3	0.11	33.90	. . .	. . .	0.3420	0.997	−0.717	1.805
		(10.60)	(21.49)		(8.55)			(2.28)			
Cold-rolled sheet	−61.91	1.064	301.4	0.11	35.86	−24.31	. . .	0.6521	0.997	−0.468	1.837
		(9.75)	(16.82)		(7.01)	(0.83)		(1.98)			
Cold-rolled sheet	−12.13	1.072	289.2	0.11	35.86	27.20	. . .	. . .	0.996	−0.271	2.139
		(7.96)	(13.22)		(6.22)	(1.34)					
Cold-rolled sheet	−51.85	1.104	293.9	0.11	34.72	. . .	. . .	0.4362	0.997	−0.351	1.904
		(10.61)	(17.30)		(7.47)			(2.32)			
Bars	−89.01	0.4683	201.6	0.09	98.85	−6.705	. . .	1.058	0.992	−0.786	2.683
		(3.09)	(15.50)		(12.67)	(0.16)		(2.14)			

Bars	5.099	0.4265 (2.29)	193.4 (12.57)	0.09	100.2 (10.61)	68.89 (2.28)	. . .	. . .	0.989	−0.693	2.689
Bars	−83.79	0.4762 (3.53)	200.6 (18.29)	0.09	98.49 (13.94)	. . .	. . .	0.9944 (3.70)	0.993	−0.783	2.691
Structurals	−66.09	1.018 (7.23)	161.5 (13.52)	0.09	64.76 (9.76)	. . .	63.34 (3.47)	0.7195 (2.97)	0.995	−0.321	2.165
Structurals	6.897	0.7427 (4.88)	168.3 (9.94)	0.09	73.82 (9.41)	. . .	64.62 (2.54)	. . .	0.991	−0.251	2.258
Structurals	−61.11	0.8597 (4.55)	170.8 (10.56)	0.09	75.97 (8.50)	. . .	. . .	0.7270 (2.03)	0.989	−0.432	2.028
Plate	−26.62	1.373 (9.16)	181.8 (11.26)	0.10	47.91 (5.75)	. . .	61.97 (3.03)	−0.01359 (0.05)	0.992	−0.790	2.722
Plate	−30.19	1.370 (7.29)	187.6 (6.55)	0.10	39.60 (5.05)	. . .	96.14 (3.50)	. . .	0.989	0	2.546
Plate	−22.84	1.236 (6.20)	191.2 (8.65)	0.10	58.97 (5.67)	. . .	. . .	0.0002 (0.00)	0.986	−0.766	2.347

Note: Figures in parentheses are t-statistics.
a. Hildreth-Liu correction for serial correlation.

Table B-2. *Effect of Japanese Consumption on Japanese Export Prices*

Product and period	*Constant*	*JMAT-COST*	*JLABOR-COST*	*g*	*D62*	*D74*	*DEV-TREND*	*DEV-TREND$_{t-1}$*	*CON/CAP*[a]	*CUJAP*	R^2	ρ^b	*Durbin-Watson statistic*
Hot-rolled sheet,	14.59	0.8313	214.8	0.11	−24.72	33.47	42.18	. . .	−16.82	. . .	0.996	0.534	2.236
1957–76		(14.09)	(17.96)		(12.90)	(9.34)	(3.38)		(1.62)				
Cold-rolled sheet,	30.18	0.8257	167.96	0.09	−30.65	45.58	. . .	. . .	−65.46	0.7244	0.968	−0.277	2.111
1957–76		(3.86)	(6.17)		(3.80)	(4.04)			(1.03)	(0.91)			
Bars, 1957–76	−27.57	0.4019	211.0		0.09	97.52	52.05	. . .	42.59	. . .	0.991	−0.668	2.554
		(4.37)	(18.32)			(15.59)	(2.94)		(3.27)				
Structurals,	−16.64	0.8312	174.52	0.09	−6.42	70.80	. . .	61.13	35.74	. . .	0.994	−0.358	2.244
1958–76		(7.58)	(7.58)		(2.31)	(11.83)		(3.86)	(2.64)				
Plate, 1958–76	−35.54	1.324	193.95	0.10	. . .	48.19	. . .	54.63	9.461	. . .	0.987	−0.615	2.512
		(10.63)	(10.09)			(5.48)		(2.57)	(0.58)				

Note: Figures in parentheses are *t*-statistics.
a. CON/CAP = Japanese domestic steel consumption divided by Japanese raw steel capacity.
b. Hildreth-Liu correction for serial correlation.

Table B-3. *Determinants of U.S. Producer Prices, 1962–76*

Product	*Constant*	*USMAT-COST*	*USLABOR-COST*	*g*	*D69*	*D70*	*D71*	*USIM-PRICET*	$CUUS_{t-1}$	R^2	ρ^a	*Durbin-Watson statistic*
Hot-rolled sheet	50.84	1.001	1.087	0.01	−3.599	−3.935	. . .	0.1224	0.09532	0.999	−0.960	3.175
		(37.68)	(2.47)		(3.96)	(4.41)		(11.00)	(4.26)			
Cold-rolled sheet	45.98	0.9678	7.255	0.01	4.498	. . .	2.89	0.1046	0.1043	1.000	−0.961	2.965
		(30.52)	(14.15)		(4.74)		(2.94)	(10.63)	(4.67)			
Bars	9.101	0.7700	18.49	0.02	−19.85	−6.437	. . .	0.08732	0.3245	0.998	−0.687	2.138
		(7.74)	(8.75)		(7.22)	(2.26)		(3.47)	(4.09)			
Structurals	2.165	0.7302	20.76	0.02	. . .	. . .	8.409	0.05002	0.1375	0.999	−0.598	2.477
		(9.80)	(12.92)				(4.33)	(2.70)	(2.80)			
Plate	2.514	0.5421	20.36	0.01	. . .	. . .	4.203	0.07391	0.1581	0.999	−0.442	2.391
		(7.61)	(18.46)				(2.12)	(4.74)	(3.26)			

Note: Figures in parentheses are *t*-statistics.

a. Hildreth-Liu corrrection for serial correlation.

Table B-4. *Estimated Domestic Demand Elasticities for Finished Steel Products*[a]

Product and period[b]	*Domestic price (P_D)*	*Import price (P_M)*	*Industrial production (Y)*
Hot-rolled sheet			
1959–76	−0.54[c]	0.31	1.24
	(1.74)	(1.62)	(6.99)
Cold-rolled sheet			
1957–76 *(OLS)*	−0.62	0.31	0.72
	(1.88)	(1.58)	(4.00)
1956–76 *(TSLS)*	−1.49	0.77	1.14
	(3.26)	(3.12)	(4.84)
Bars			
1957–76 *(OLS)*	−0.83	0.39	0.63
	(3.07)	(2.38)	(3.90)
1956–76 *(TSLS)*	−1.30	0.63	0.72
	(6.13)	(5.60)	(7.29)
Structurals			
1957–76 *(OLS)*	−1.99	1.08	0.96
	(7.69)	(6.57)	(6.51)
1956–76 *(TSLS)*	−1.98	1.01	1.07
	(8.93)	(7.72)	(8.72)
Plate			
1957–76 *(OLS)*	−1.39	0.92	0.81
	(5.18)	(5.76)	(3.90)
1956–76 *(TSLS)*	−1.81	1.09	1.12
	(8.82)	(9.37)	(8.13)

a. The figures in parentheses are *t*-statistics. The estimated equations are

$$Q_D = a_o P_D^{a_1} P_M^{a_2} Y^{a_3}$$

$$Q_S = \beta_o\ P_D^{\beta_1} CUUS^{\beta_2}$$

where
$Q_S = QD$ = domestically supplied finished steel shipments (annual date),
QM = imports of finished steel,
$CUUS$ = capacity utilization in the United States,
P_D = domestic price realizations,
P_M = import price (including freight and tariff),
Y = industrial production index (Federal Reserve Board).

b. *OLS* = ordinary least squares, and *OLS* estimates are corrected for serial correlation by the Hildreth-Liu method. *TSLS* = two-staged least squares.

c. In this equation the domestic price is lagged one year.

Table B-5. *Estimated Import-Demand Elasticities*[a]

Product and period	*Import price (P_M)*	*Domestic price (P_D)*	*Industrial production (Y)*
Hot-rolled sheet			
1957–76	−10.45[b]	14.41[b]	3.38
	(7.28)	(5.42)	(3.49)
1962–76	−2.54[b]	. . .	7.16
	(3.88)		(4.52)
Cold-rolled sheet			
1958–76	−7.35[c]	9.78[c]	4.50
	(3.02)	(1.90)	(1.81)
1962–76	−5.17	7.29	2.41
	(4.57)	(3.09)	(1.94)
Bars			
1958–76	−1.30[b]	. . .	3.88
	(5.06)		(12.14)
1962–76	−1.09[b]	. . .	3.63
	(4.94)		(8.13)
Structurals			
1958–76	−1.42 −1.84[b]	4.11	. . .
	(3.63) (5.71)	(8.57)	
1962–76	−0.54		2.01
	(3.50)		(6.88)
Plate			
1957–76	−2.82	3.50	2.51
	(2.99)	(2.01)	(2.27)
1962–76	−1.09	. . .	4.58
	(2.63)		(10.26)

a. The figures in parentheses are *t*-statistics. The estimated equation is

$$Q_M = a_o P_M^{a_1} P_D^{a_2} Y^{a_3}$$

where

Q_M = imports of finished steel (annual date),
P_M = import price (including freight and tariff),
P_D = domestic price realizations,
Y = industrial production index (Federal Reserve Board).

All estimates are obtained by ordinary least squares.

b. Independent variables lagged one year.

c. Independent variables lagged two years.

Index